Oryx Frontiers of Science Series

RECENT ADVANCES AND ISSUES IN PHYSICS

Oryx Frontiers of Science Series

Chemistry
Recent Advances and Issues in Physics
Recent Advances and Issues in Environmental Science
Recent Advances and Issues in Biology
Recent Advances and Issues in Computers
Recent Advances and Issues in Anthropology
Recent Advances and Issues in Geology
Recent Advances and Issues in Meteorology
Recent Advances and Issues in Astronomy

Each new volume in this series is the ideal first-stop source
for information on cutting-edge issues in the field.

Series features include

Recent research • Current social and ethical issues
New technology and applications • Biographies of key individuals
Documents, speeches, statements, and reports
Unsolved questions and research trends • Career opportunities
Organizations and associations • Print and nonprint resources
Glossary of terms • Comprehensive subject index

Oryx Frontiers of Science Series

RECENT ADVANCES AND ISSUES IN PHYSICS

by David E. Newton

Oryx Press
2000

The rare Arabian Oryx is believed to have inspired the myth of the unicorn. This desert antelope became virtually extinct in the early 1960s. At that time, several groups of international conservationists arranged to have nine animals sent to the Phoenix Zoo to be the nucleus of a captive breeding herd. Today, the Oryx population is over 1,000, and over 500 have been returned to the Middle East.

© 2000 by David E. Newton

Published by The Oryx Press
4041 North Central at Indian School Road
Phoenix, Arizona 85012-3397
http://www.oryxpress.com

Cover images of vortex dynamics courtesy of
Professor C. Fred Driscoll, Department of Physics,
University of California, San Diego. See also pages 47–49.

Published simultaneously in Canada
Printed and bound in the United States of America

∞ The paper used in this publication meets the minimum requirements of
American National Standard for Information Science—Permanence
of Paper for Printed Library Materials, ANSI Z39.48, 1984.

Library of Congress Cataloging-in-Publication Data

Newton, David E.
  Recent advances and issues in physics / by David E.
Newton.
    p. cm. — (Oryx frontiers of science series)
  Includes bibliographical references and index.
  ISBN 1-57356-171-1 (alk. paper)
  1. Physics. I. Title. II. Series.
  QC21.2.N52   1999 2000
  530–dc21   99-045843
                                             CIP

To Schuyler ("Ty") Slater
Mentor, Colleague, and Friend

With appreciation for your
patience, understanding, and leadership.

CONTENTS

PREFACE

What's new in physics? Answering that question is no small task. Research in physics seems to accelerate each year, with new discoveries and inventions being reported at a staggering rate. Indeed, in a book of this size, recent developments in physics can be reviewed only by making the kinds of choices that are not easy. Is it more important to discuss progress in the search for fundamental particles, or to describe developments in nanotechnology? Should equal attention be paid to progress in astrophysics and in geophysics? Are some new announcements inherently more important or more significant than others?

The research reported in this book will not unanimously be acknowledged as that most worthy of attention. It should, nevertheless, provide readers with a taste of the developments taking place in physics today. Chapters 1 and 2 are devoted to advances in basic and applied physics research that have been reported within recent years.

Chapter 3 is devoted to a discussion of unsolved problems in physics. That such a chapter should exist is something of an anachronism. It suggests that scientific research consists of beginnings and endings: one project comes to an end, and scientists begin to think about what they should work on in the future. Such is not the case, of course. Research should be compared, instead, to a seamless string or web in which old problems are constantly being solved, leading to new studies of related issues. The topics listed in Chapter 3 are, however, among those that many physicists regard as of critical interest for the future.

Chapter 4 provides brief biographical sketches of individuals whose research is discussed in Chapters 1 and 2 or who made seminal contributions to those fields of research. The purpose of these sketches is to provide an overview of the kinds of backgrounds, interests, and experiences physicists have.

Progress in any field of science, physics included, does not take place in a vacuum. It almost always has some impact on society and, in turn, it is affected by social factors such as economic, philosophical, and ethical issues. Even in fields as abstruse as particle physics, the direction of research is influenced by how much money the federal government is willing to spend on such projects and to what extent the general public sees those projects as having value. Some of the ways in which physics and society as a whole have impact on each other is the subject of Chapter 5.

Some important documents relating to topics raised in Chapter 5 are provided in Chapter 6. These documents are presented in order that readers may see what proponents have had to say about various positions on these topics and what legal and/or formal actions have been taken on them.

Chapter 7 provides an overview of career opportunities in physics. The reader is introduced to the kinds of jobs available to physicists, the kinds of training required for such jobs, the opportunities and rewards to be expected in these fields, and print and electronic resources from which additional information can be obtained.

Chapter 8 summarizes some general statistical information about physics that may be of interest to the general reader. Included among those data are budgetary allocations for various fields of physical research and education and career patterns among physicists.

Chapter 9 contains a list of organizations and associations concerned with physics. These organizations and associations range from large, general-interest groups, such as the American Institute of Physics, to smaller, more specific groups, such as the Society of Rheology.

A list of print and electronic sources from which further information about physics can be obtained is provided in Chapter 10. Because this book deals primarily with recent advances in physics, the list focuses on journals and electronic sources in which current information can be found.

Finally, Chapter 11 provides a glossary of terms used in the book that may be unfamiliar to the general reader.

Oryx Frontiers of Science Series

RECENT ADVANCES AND ISSUES IN PHYSICS

CHAPTER ONE
Basic Research in Physics

INTRODUCTION

Scientific research is commonly divided into two general categories: basic and applied. Basic research is carried out primarily to learn more about the natural world. Applied research is conducted to solve some specific problem.

This division is somewhat arbitrary and does not necessarily correspond to the realities of scientific work. Many research projects do focus on gaining fundamental knowledge of the universe, although they also may result in the invention of some device that will have practical use in everyday life.

Accepting this ambiguity in categorization, this chapter focuses on progress in basic research in physics reported during the late 1990s. No list of worthwhile projects could be considered complete or adequate. However, the studies reported in this chapter are indicators of the kinds of progress that physicists in a variety of fields have been making.

PARTICLE PHYSICS

One of the oldest and most basic questions in science concerns *fundamental particles*. What is it out of which everything else is made?

For most of the nineteenth century, scientists believed the answer to that question was *atoms*. Just as the 1800s drew to a close, however, British physicist J. J. Thomson discovered a particle smaller than the atom, the electron. Thomson's research confirmed that atoms are not fundamental particles. Instead, they are themselves composed of even smaller particles. The electron was the first of these *subatomic* particles to be found.

For much of the twentieth century, most high school science students have been taught that matter consists of atoms that, in turn, are made of three smaller subatomic particles: protons, neutrons, and electrons.

Even as this model was being taught in schools, however, new discoveries were being made indicating that this view of matter is greatly oversimplified. Largely as the result of experiments carried out in large-particle accelerators ("atom smashers"), scientists had, by the early 1960s, discovered dozens of additional "fundamental" particles. These particles were given names such as muons, pions, K particles, lambda particles, upsilon particles, and neutrinos.

Few scientists believed that all of these particles were truly fundamental. They began to look for some simplifying scheme that would make sense out of the "particle zoo" that had been discovered. The outlines for such a scheme were suggested by American physicist Murray Gell-Mann in the 1950s. The theory that began with Gell-Mann's work is now known as the Standard Model.

According to the Standard Model, only two types of particles are truly fundamental. Those particles are known as *quarks* and *leptons*. The table on page 3 outlines the main features of the Standard Model.

As the table shows, quarks and leptons occur in pairs and can be detected at various energy levels. Energy level 1, for example, represents the everyday conditions under which the world normally operates. Energy level 2, by contrast, represents much higher levels of energy that are encountered only in particle accelerators and high-energy astronomical events, such as cosmic ray showers. Energy level 3 represents extraordinarily high energy levels that probably existed in the natural world only once in the history of the universe: during its creation.

What the table suggests is that only one pair of quarks and one pair of leptons exist at normal energy levels. One can observe the other pairs of quarks and leptons only at much higher energy levels.

According to the Standard Model, quarks can combine with each other to form more complex particles. A proton, for example, consists of two up quarks combined with one down quark. It can be represented as: uud. The particle known as sigma zero (Σ^0) consists of one up quark, one down quark, and one strange quark (uds). The sigma zero particle, because it

Energy Level	Quarks		Leptons		Conditions under Which the Particles Form
	Name	Symbol	Name	Symbol	
1	top (truth)	t	tau	τ	Creation of the universe; most powerful particle accelerators
	bottom (beauty)	b	tau neutrino	ν_τ	
2	strange	s	muon	μ	Cosmic ray events; most particle accelerators
	charm	c	muon neutrino	ν_μ	
3	up	u	electron	ε	Everyday events
	down	d	electron neutrino	ν_ε	

The Standard Model: Fundamental Particles

contains a strange quark, does not occur under normal conditions, but can be produced only in particle accelerators or other sources of high energy.

The Standard Model has been expanded considerably in recent years to accommodate other types of particles found in experiments or predicted by theory. It is too complex to review in detail in this book, although this basic background will assist the reader in understanding some of the important discoveries discussed below.

Discovery of the Top Quark

Research on quarks in the third level requires the most powerful accelerators in the world. It is only in the energy released in these accelerators that the top and bottom quarks can ever be expected to appear. In 1977, researchers at the Fermi National Accelerator Laboratory (Fermilab), outside Chicago, announced the discovery of the bottom quark. That discovery was made by firing a beam of protons into a beam of antiprotons. Antiprotons are identical to protons except for their charge: They carry a negative charge rather than a positive charge. The energy released in the proton-antiproton collisions was sufficient to briefly allow the formation of the bottom quark.

The discovery of the bottom quark confirmed that a third level of quarks did, indeed, exist. It inspired researchers to begin work almost immediately on experiments to find the last remaining quark, the top quark.

That search went on for nearly 20 years. Along the way, a number of premature claims were made that could not be substantiated. In 1985, for example, researchers at the Centre Européen pour la Recherche Nucléaire (CERN) announced that they had produced a top quark. Other scientists were unable to confirm that report. Similar announcements by Fermilab physicists in 1992 also turned out to be false alarms.

Finally, apparently solid evidence for the top quark was announced by Fermilab researchers in April 1994. After analyzing more than 16 million proton-antiproton collisions, Fermilab physicists reported finding evidence of a handful of top quarks. They continued to refine their data, and a year later, they reported solid evidence for the creation of about 75 top quarks. Their results have been accepted by most physicists around the world as confirming the discovery of the last quark.

It is hardly surprising that the top quark was such difficult prey. In the first place, it is massive as far as subatomic particles are concerned. The mass of a subatomic particle is typically measured in its energy equivalent; in these terms, the top quark has a mass of about 180 GeV (billion electron volts). This mass is roughly similar to that of a single gold atom. The total mass of the top quark is confined to a very small region, however, with a diameter of no more than 10^{-18} m.

One problem in creating a top quark, then, was to generate enough energy for a particle of this mass to appear. That energy is available under only rare circumstances in the proton-antiproton collisions generated at the Fermilab Tevatron (particle accelerator).

Another problem complicating the search for the top quark was its very short lifetime. Theorists predicted that the top quark would survive no more than about 0.4×10^{-24} seconds before breaking apart into other subatomic particles.

It was, in fact, the decay scheme of the top quark that allowed physicists to recognize its existence. When the top quark decays, it breaks down into a bottom quark and a W boson (another type of subatomic particle). Both the bottom quark and the W boson are themselves unstable. They break down according to well-known decay schemes. It was the pattern of decay schemes—top quark to bottom quark and W boson; bottom quark and W boson into simpler particles—that physicists searched for in their experiments. More than 450 physicists were involved in this search. The instrument they used, the Fermilab Collider Detector, is one of the most sophisticated and sensitive detection devices ever constructed for use with particle accelerators.

Scientists believe that top quarks were formed within an instant of the creation of the universe following the Big Bang. All of the top quarks formed at that instant must have decayed in a vanishingly short period of time to form bottom quarks and W bosons. These particles decayed, in turn, to form other fundamental particles that are more stable at the lower energy levels that appeared in the minutes and hours following the Big Bang.

References

"Discovery of the Top Quark" <http://www.fnal.gov/pub/top95/ top95_background.html> accessed 30 April 1999.

"Discovery of the Top Quark" <http://lutece.fnal.gov/Papers/PhysNews95.html> accessed 30 April 1999.

Liss, Tony M., and Paul L. Tipton, "The Discovery of the Top Quark," *Scientific American*, September 1997; also <http://www.sciam.com/0997issue/ 0997tipton.html> accessed 30 April 1999.

Yarris, Lynn, "First Experimental Evidence for Top Quark Announced," <http:// www.lbl.gov/Science-Articles/Archive/top-quark-first-evidence.html> accessed 30 April 1999.

Discovery of the Last Normal Meson

The name *meson* comes from a term originally suggested by the Japanese physicist Hideki Yukawa (1907–81) in the 1930s. Yukawa theorized that a particle existed with the ability to hold protons and neutrons together in the nucleus. He calculated that the mass of such a particle would be midway between that of a proton and an electron. When a particle meeting that description was later discovered, it was called a *mesotron* (or "medium-mass-tron"). Although the particle later turned out not to be the one Yukawa had predicted, its name was retained and shortened to *meson*.

Over the next half century, a variety of mesons were discovered. These mesons differed from each other in their mass and their lifetimes. They were given names such as the mu meson, the pi me-

Chart showing possible quark-antiquark combinations. Each box includes an example of a meson containing that combination. The black square highlights the last-discovered meson combination, made of a charm quark and an anti-b. *Courtesy of FermiNews.*

$u\bar{u}$ π°,η,η'	$u\bar{d}$ π^+	$u\bar{s}$ K^+	$u\bar{c}$ $\overline{D^\circ}$	$u\bar{b}$ B^+
	$d\bar{d}$ π°,η,η'	$d\bar{s}$ K°	$d\bar{c}$ D^-	$d\bar{b}$ B°
		$s\bar{s}$ η,η'	$s\bar{c}$ D_s^-	$s\bar{b}$ B_s°
			$c\bar{c}$ J/ψ	$c\bar{b}$ B_c^+
				$b\bar{b}$ Υ

son, the K meson (also known as a *kaon*), the rho meson, and the omega meson. Today, physicists know that all mesons have one important characteristic in common: They all consist of one quark and one anti-quark. For example, one of the K mesons, the K^+ meson, consists of an up quark combined with an anti-strange quark (designated as $\overline{s}u$).

Only five of the six quarks can form mesons. The top quark has such a short lifetime that it cannot form a meson. The five other quarks and their corresponding antiparticles can form a total of 15 mesons.

Until recently, only 14 of these 15 mesons had been found. The last remaining meson is one formed by the combination of a bottom quark and a charm quark. Then, in April 1998, researchers at the Fermi National Accelerator Laboratory (Fermilab) announced the discovery of the last meson. They had searched through more than 100 million collisions between protons and antiprotons observed at the Fermilab Tevatron accelerator, and among those collisions were 19 that appeared to meet the criteria for a bottom quark–charm meson decay. During such a decay, other subatomic particles are formed that travel with distinctive pathways in distinctive time periods through a detector. The Fermilab's discovery appears to complete the set of all predicted "normal" mesons. An additional set of "exotic" mesons is thought to exist and is currently being studied.

References

Kestenbaum, David, "Physicists Find the Last of the Mesons," *Science*, 3 April 1998: 35.
"Last of the Normal Mesons," *Science News*, 25 April 1998: 271.

Exotic Mesons

At times, there seems to be no end to the variety of subatomic particles physicists can produce. In September 1997, physicists at the Brookhaven National Laboratory in New York reported the discovery of yet another new type of particle: an exotic meson. An exotic meson is a subatomic particle that consists of a quark, an antiquark, and a gluon. By compari-son, a normal meson consists of a quark paired with an antiquark, and baryons (such as protons, electrons, and neutrons) consist of three quarks.

The third particle in an exotic meson, the gluon, is a force-carrying particle that usually exists in a virtual state. That is, it behaves as a bundle of energy rather than as a concrete particle. For example, gluons are thought to exert the force that holds protons and neutrons together in an atomic nucleus. On rare occasions, however, a gluon may take on just the right amount of energy to manifest itself as a particle. Its existence as a particle in an exotic meson is one such occasion.

In the Brookhaven experiment, a form of mesons known as pions were accelerated to nearly the speed of light in the laboratory's Alternating Gradient Synchrotron. The beam of pions was then allowed to collide with a target of liquid hydrogen. Researchers studied more than 200 million collisions between pions from the beam and protons of the liquid hydrogen. They found about 40,000 events that could be interpreted as resulting in the formation of an exotic meson.

Brookhaven physicists explained their results as follows. The pion consists of a down quark and an anti–up quark held together by a gluon in its virtual state. The gluon can be thought of as a wire joining the two particles. Energy released by the pion-proton collisions caused the gluon to vibrate at a high frequency, just as a guitar string vibrates when it is plucked. At just the right frequency, the gluon's energy appeared as mass. It was, for one brief moment, a real particle.

The gluon did not remain in this state for very long. In fact, the exotic meson has a lifetime of only 10^{-23} seconds. It is obvious that researchers did not observe the exotic meson itself; instead, they identified the particle on the basis of the charge, energy, and direction taken by particles formed during its decay. These characteristics corresponded to the properties that theorists had predicted for those of an exotic meson.

The formation of a quark-antiquark-gluon exotic meson is not the only possible explanation of the results of the Brookhaven experiment. Theorists have predicted at least three types of exotic mesons. One type is a *hybrid* represented by the quark-antiquark-gluon structure. A second type of exotic meson is called a *diquarkonium* and consists of two pairs of quarks (two quarks and two antiquarks). A third type of exotic meson is called a *glueball* and consists of two or three gluons bound to each other. It is conceivable that the Brookhaven experiment resulted in the formation not of a hybrid meson, but of a diquarkonium.

Reports like those of the Brookhaven group are subjected to very close scrutiny. Such results are regarded as tentative until other researchers are able to produce comparable results in their own laboratories. In the case of the exotic meson, confirmatory reports appeared almost immediately from physicists working at the Centre Européen pour la Recherche Nucléaire (CERN), and the Serphukov Laboratory in Russia reported similar findings.

Interestingly enough, however, the discovery of an exotic meson announced by Brookhaven physicists was questioned by members of the Brookhaven team itself. That team consisted of 51 physicists from 7 institutions. But 10 members of the team refused to sign the paper announcing the discovery. They indicated that they were not completely convinced that their results proved the existence of an exotic meson—they felt that those results might be explained in other ways.

Physicists elsewhere have also raised questions about the Brookhaven results. They point out that those results differ slightly from the findings reported by CERN and Serphukov. It is still not clear whether the small differences among reports are the result of experimental errors or whether they represent different types of exotic mesons or some entirely different kind of particle.

References

Napolitano, Jo, "Dissent on Find at Brookhaven," *Newsday*, 4 September 1997: A27.

Peterson, I., "Exotic Needle Found in Particle Haystack," *Science News*, 6 September 1997: 148.

"Physics of Experiment 852" <http://lemond.phy.bnl.gov/intro/physics.html> accessed 30 May 1999.

"Scientists Detect Evidence of a Rare, Elusive New Subatomic Particle," *Los Angeles Times*, 1 September 1997: A11.

Taubes, Gary, "New Exotic Particle Points to Double Life for Gluons," *Science*, 12 September 1997: 1609.

Neutrino Mass

Perhaps the most exciting news in particle physics in the 1990s was announced in June 1998. A group of 120 physicists from 23 institutions in Japan and the United States reported that neutrinos may have mass. This news was met with comments such as "a huge step forward," "incredibly impressive," and "one of the most thrilling moments of my life."

Neutrinos have long been one of the most puzzling and interesting of all subatomic particles. They were first hypothesized in 1930 by the Italian physicist Enrico Fermi. Fermi chose the name *neutrino* for the hypothesized particle—"little neutral one" in Italian—because it carries no electrical charge.

Neutrinos later became an essential part of the Standard Model (see table on page 3). That theory requires the existence of three kinds of neutrinos: the electron neutrino, muon neutrino, and tau neutrino.

For nearly seven decades, most physicists have believed that neutrinos have no mass. One reason for that belief is that untold numbers of neutrinos exist in the universe, but they are virtually impossible to detect. For example, trillions of neutrinos pass through your body every second, but you are completely unaware of their presence. They do not interact with matter in any way. Particles with mass would not behave in that way.

From the start, however, some physicists have rejected the concept of a totally massless neutrino. They have devised a number of experiments to prove that neutrinos do have some mass, even if it is only a vanishingly small quantity. Until 1998, none of those experiments produced convincing evidence to support this hypothesis.

The success of the Japanese-U.S. experiment was partly due to the very large and sophisticated equipment with which it was conducted. That equipment, the Super-Kamiokande (or Super-K) detector, contained the largest neutrino-detecting equipment ever built. It was constructed at the bottom of a zinc mine and consisted of a 50,000-ton tank of ultrapure water surrounded by 13,000 photomultiplier tubes. These tubes are able to photograph the light produced any time a neutrino collides with a molecule of water, an event that occurs only very rarely.

The Super-K detector was oriented to count collisions produced from neutrinos entering at the top of the mine (the detector was situated about a kilometer beneath the Earth's surface), and from the bottom of the mine. Neutrinos entering the bottom of the mine would first have passed completely through the Earth, a distance of some 13,000 kilometers. The basic discovery made by the Super-K research team was that the number of electron neutrinos entering the detector was the same from either direction. But the number of muon neutrinos from the bottom of the

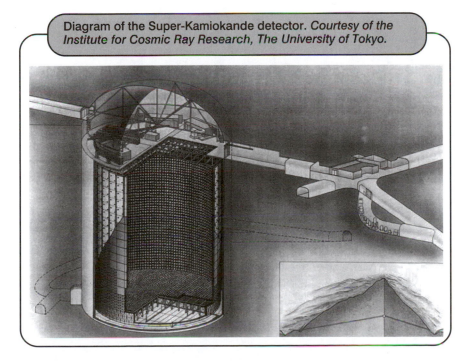

Diagram of the Super-Kamiokande detector. *Courtesy of the Institute for Cosmic Ray Research, The University of Tokyo.*

detector was about half (139) of those entering from the top of the detector (256).

The interior of the Super-Kamiokande detector. *Courtesy of the Institute for Cosmic Ray Research, The University of Tokyo.*

The conclusion to be drawn from this finding is fairly straightforward: Muon neutrinos passing through the Earth had a greater time to "disappear" than those that traveled only a kilometer to the top of the detector. The question was how those neutrinos could have "disappeared."

One way to answer this question is to assign neutrinos a very special property known as *oscillation*. Oscillation refers to the concept that the three forms of neutrinos are not really fundamentally different, but that any one type of neutrino has the ability to be transformed into any other type of neutrino. As a neutrino travels through space, then, it may appear to be an electron neutrino at one moment, a muon neutrino at another moment, and a tau neutrino at yet another moment.

This concept is easier to understand if one thinks of a neutrino as traveling in a wave, in much the same way that light travels. The neutrino, however, may consist of three different kinds of waves all traveling together. At any one moment, one kind of wave may predominate, while at another moment, another kind may predominate.

The most important point about oscillation, however, is that this behavior can occur only with particles that have mass. The observed differences in the numbers of muon neutrinos in the Super-K experiment suggest that some muon neutrinos were transformed into (unobservable) tau neutrinos as they passed through the Earth. Such oscillation can occur, however, only if one or another form of the neutrino has mass. It is not possible to determine *which* type of neutrino has mass, only that one form must have this property.

The decay patterns produced when neutrinos strike water molecules in the Super-K detector make it possible to estimate the mass of a neutrino. At this point, the best estimate is that the neutrino mass is about 0.07 ev (electron volt), which is about 0.0000001 the mass of the electron.

The mass of a single neutrino is obviously almost insignificant. What is important, however, is the total mass of *all* neutrinos in the universe. Given the vast number of neutrinos, that total mass may be very great indeed. Some physicists estimate that even a mass of 0.07 ev for a single neutrino could result in a total mass of all neutrinos equivalent to the known mass of the universe. If that were true, neutrino mass could become a very important factor in predicting the future of the universe itself.

References
Brainard, J., "Ghostlike Particles Carry a Little Weight," *Science News*, 13 June 1998: 374.
Dawson, Jim, "Mass Distinction," *Minneapolis Star Tribune*, 24 June 1998: 16A.
Gibbs, W. Wayt, "A Massive Discovery," *Scientific American*, August 1998: 18–19.

Kestenbaum, David, "Neutrinos Throw Their Weight Around," *Science*, 11 September 1998: 1594–95.

Maugh, Thomas H., II, "World of Physics Jolted by Finding on Neutrinos," *Los Angeles Times*, 5 June 1998: A1.

Normile, Dennis, "Weighing in on Neutrino Mass," *Science*, 12 June 1998: 1689–90.

Evidence for the Tau Neutrino

Only a month after the Super-K results were announced (see **Neutrino Mass**, above), another breakthrough in neutrino research was announced. Physicists at the Fermi National Accelerator Laboratory (Fermilab) reported that they may have seen the first signs of the tau neutrino. The tau neutrino is the only one of the three neutrinos that has never been observed by physicists.

The tau neutrino presents a problem for researchers at least partly because of the properties it shares with other neutrinos. It travels at close to the speed of light and interacts with matter only in rare circumstances. And it poses one other challenge for researchers. It exists exclusively at very high energy levels, such as those produced in the most powerful particle accelerators. Overall, the chances of observing a tau neutrino have been exceedingly small.

The Fermilab experiment involved the collision of a beam of protons traveling at close to the speed of light with a target of tungsten metal. The collision between protons and tungsten atoms resulted in the formation of a handful of tau neutrinos. Those neutrinos were observed with a detector consisting of piles of photograph plates coated with silver bromide. The presence of a tau neutrino was observable on these plates because of the tau particle it produces when it decays.

Over a period of four months, the Fermilab team found three characteristic tau neutrino tracks. When they announced this discovery, they also explained that they planned to continue their analysis of the proton-tungsten collisions until they could identify seven more tau neutrino tracks. At that point, they would have even greater confidence that they had finally trapped the most elusive of the neutrino family.

Reference

Antia, Meher, "First Glimpse of the Last Neutrino?" *Science*, 10 July 1988: 155.

Further Reading

Gribbin, John R., *Case of the Missing Neutrinos: And Other Curious Phenomena of the Universe*, New York: Fromm International, 1998.

Gribbin, John R., and Mary Gribbin, eds., *Q Is for Quantum: An Encyclopedia of Particle Physics*, New York: The Free Press, 1999.

Kane, Gordon, and Heather Mimnaugh, eds., *The Particle Garden: Our Universe As Understood by Particle Physicists*, Reading, MA: Perseus Press, 1996.

Schwarz, Cindy, *A Tour of the Subatomic Zoo: A Guide to Particle Physics*, College Park, MD: American Institute of Physics, 1996.

Solomey, Nickolas, *The Elusive Neutrino: A Subatomic Detective Story*, New York: W. H. Freeman, 1997.

Stwertka, Albert, *The World of Atoms and Quarks*, New York: Twenty First Century Books, 1995.

Trefil, James, *From Atoms to Quarks: An Introduction to the Strange World of Particle Physics*, New York: Anchor Books, 1994.

ASTROPHYSICS

One of the most intriguing facts about research in physics today is the relationship between studies of particle physics and cosmology. Cosmology is the study of the origin of the universe, its present structure and properties, and its ultimate fate. The discovery that physicists have made is that their research on the smallest particles known, such as quarks and leptons, provides critical information about the largest structures known— galaxies, galactic structures, and the universe itself. And vice versa. Each new discovery about the universe provides new information or raises new questions about the fundamental particles of which matter is made.

The last two decades have been an era of unprecedented discoveries in astronomy and astrophysics. These discoveries have come about, to a large extent, because of the availability of powerful new detection tools, such as the Keck Telescope on top of Mauna Kea in Hawaii, the Hubble Space Telescope, and other specialized observational satellites, such as NASA's Cosmic Background Explorer (COBE). These devices have provided scientists with a more detailed view of a greater variety of objects, and objects at greater distance from the Earth, than had even been imagined as possible 30 years ago.

Inflation Theory

In a fundamental sense, more and better information is good, of course, but many of the discoveries have created new problems for fundamental theories of cosmology. In some cases, astrophysicists are faced with the question of whether some of the massive quantity of data they are receiving is incorrect, or whether theoretical explanations on which they have come to depend are wrong. The dawn of the twenty-first century is clearly one of the most confusing, frustrating, challenging, and exciting periods in all of modern astrophysics.

The basic context within which most cosmological discoveries are assessed is the inflation (or inflationary) theory. *Inflation theory* is a cosmological theory originally developed by the Russian physicist Alexei

Starobinsky in the late 1970s. At the time, officials of the then–Soviet Union prevented Russian scientists from corresponding with scientists in other parts of the world. As a result, Starobinsky's work was not well known outside the Soviet Union.

By 1981, however, the inflation theory had been "rediscovered" independently by Alan Guth, then at the Massachusetts Institute of Technology (MIT). Today, Guth is widely acknowledged as the "father" of modern inflation theory.

Inflation theory evolved out of, and refined, the early Big Bang theory of the creation of the universe. According to the Big Bang, the universe was created at some instant approximately 20 billion years ago. Inflation theory provides details of the process of creation and deals with a handful of essential experimental observations.

One of those observations is the *horizon problem*. The horizon problem means that all parts of the universe appear to be exactly alike, regardless of the direction in which one looks. The question is how all parts of the universe can have essentially the same physical characteristics when they are separated from each other by such great distances.

The second observation with which inflation theory deals is the *smoothness problem*. Stars, galaxies, and other large concentrations of matter all appear to be spread out across the universe in relatively equal concentrations.

The third observation is the *flatness problem*. One can predict three possible conditions for the universe. An *expanding universe* is one in which all objects in the universe are constantly moving away from each other. The overall density of space constantly becomes less and less. A *contracting universe* is one in which all objects in the universe eventually feel the force of gravitational attraction and begin to move inward toward each other. The ultimate fate of this kind of universe would be a *Big Crunch*, the opposite of the Big Bang, when all matter is brought back to a single point. A third alternative is known as a *flat universe*. A flat universe is one in which the rate of expansion gradually slows down over time, but it never becomes zero. Most scientists today believe that we live in a flat universe, so any theory of the origin of the universe must explain why that condition exists today.

According to Guth's inflation theory, the universe was created out of nothing as the result of random fluctuations in a vacuum. Quantum theory says that energy and matter can occur even in total emptiness, such as a perfect vacuum. That is, it is possible for a pair of particles, such as a quark and an antiquark, to appear instantaneously out of a vacuum, provided that they combine with each other immediately and disappear.

It is also possible, according to quantum theory, for bursts of energy to appear, and then disappear, within the vacuum.

Guth suggested that the Big Bang was initiated as the result of the release of energy from the primordial vacuum, followed by a force that drove that energy outward so fast that it could not return to the vacuum. The initial explosion occurred with unbelievable speed. The universe doubled in size, according to Guth, about every 10^{-34} second. Within a fraction of a second, the young universe went from a singularity (a point with no dimensions) to the size of a grapefruit. The force with which this expansion began is still pushing matter outward along the paths that we can observe today.

Guth's theory explains many of the fundamental observations made of the universe today. For example, at its very earliest stages, all parts of the universe were in intimate contact with each other. Every part "knew" what every other part was like. The horizon problem noted above can be explained on this basis.

Guth's theory has been modified and refined a number of times, especially by Henry Tye, Andrei Linde, and Paul Steinhardt. In the 1990s, many astrophysicists thought that inflation theory provided the answers to many fundamental questions of cosmology and suggested many useful hypotheses for further research. At the same time, new observations raised some difficult questions for the theory.

Dark Matter
Arguably the most serious problem with inflation theory has to do with dark matter. *Dark matter* is so called because it does not give off visible light and therefore cannot be seen with optical telescopes or the naked eye.

Astronomers have a number of reasons for believing in the existence of dark matter. For example, stars in distant galaxies revolve around the galactic center with relatively constant velocities, no matter how far they are from the center. Such a phenomenon is very different from the way planets in our Solar System revolve around the Sun. In the latter case, the orbit of planets is determined solely by the gravitational attraction of the Sun, at the center of the system. In order to account for an orbiting system in which all stars travel at the same speed, one must assume the presence of a gravitational field spread equally throughout the system.

Further evidence comes from the rate at which galaxies rotate. That rate can be predicted simply by knowing the mass of a galaxy. And the mass of a galaxy can be estimated with reasonable accuracy by measuring its luminosity, the amount of light it produces. The problem is that most galaxies rotate far more rapidly than one would expect from such calculations. One would expect the galaxies to tear themselves apart during rotation if the only mass they contain is that from luminous objects.

The explanation for these unexpected rates of revolution and rotation has traditionally been the presence of dark matter. The amount of dark matter is not, in fact, inconsequential. Various astrophysicists estimate the amount of matter "missing" in galaxies, galactic groups, and the universe itself to be anywhere from 60 to 99 percent of all the matter that actually exists in the universe. The challenge has been to find that matter.

So far, unfortunately, astronomers have been almost totally unsuccessful in finding any dark matter. There have been times when hopes were raised that progress had been made. In 1996, for example, researchers with a project known as the MACHO (for Massive Compact Halo Object) Collaboration announced that they had found large chunks of dark matter in the halo surrounding our galaxy, the Milky Way. The halo is the dim swarm of matter that surrounds the galactic disk. MACHO researchers had identified those objects by means of *gravitational lensing*.

Gravitational lensing is the bending of light from a distant object as it passes close to a massive object. The gravitational field of the massive object causes the light to come to a focus in front of the object, much as a glass or plastic prism brings light to a focus. The MACHO team identified a number of occasions on which light from a distant star suddenly increased in magnitude and then faded out again. The team hypothesized that dark matter in our galaxy's halo had acted as a gravitational lens— the dark matter had briefly intensified the light of a distant star as it passed behind the matter.

These findings did not long withstand analysis by other observers, however. One of the most serious problems arose in early 1998. MACHO researchers announced to world astronomers that a lensing effect was to be expected in mid-June 1998. Telescopes around the world were poised to photograph the effect. Measurements were designed to estimate the distance of the dark object that was to cause the lensing of the distant star's light.

The results were disappointing to the MACHO team. The measurements indicated that the massive object was much too distant to be located in the Milky Way's halo. Instead, it seemed clear that the object was actually in the Small Magellanic Cloud, the galaxy in which the distant star itself was located. Promising hints of the existence of dark matter in our own galaxy had apparently not panned out.

Additional news for dark matter theorists was reported in late 1997 by researchers at the Paris Observatory at Meudon and at the Strasbourg Observatory. These researchers had been looking for dark matter not in the halo, but in the plane of the Milky Way. These researchers (independent of each other) made use of data on the movement of stars as reported by the Hipparcos astronomical satellite. From the orbits of these stars, they were able to estimate the mass density in the galactic disk. The mass density they found matches very closely the value determined by other methods, such as the amount of light produced by visible matter in the galactic disk. Both researchers concluded that there is no mass left unaccounted for in the disk.

By the end of 1998, then, hopes for finding dark matter in our own galaxy had largely been dashed. These studies do not, of course, put the issue of dark matter to rest forever. There are far too many other places in the universe to search for this "missing" matter. However, it is somewhat discouraging not to have found any dark matter in the portion of the universe that is closest and most accessible to us and in which, therefore, it might have first been identified.

Type Ia Supernovae

One of the great benefits of the Hubble Space Telescope is its ability to see more distant objects than anything ever observed from the Earth's surface. The Hubble makes its observations, of course, far beyond any type of matter that blocks the view of Earth-based telescopes. As a result, astrophysicists are now able to look at objects billions of light-years away.

That fact also means that they can look back billions of years into the history of the universe. Light collected by the Hubble from a galaxy six billion light-years away, for example, was generated six billion years ago—about a half or a third of the time since the universe was created. Such information provides astrophysicists with a view of the way the universe looked much earlier in its history. That view can be compared with the predictions of inflation theory to see how accurately that theory has predicted cosmological events.

Findings reported in late 1997 and early 1998 have created some problems for inflation theory. According to that theory, just enough matter should have been created in the Big Bang to form a universe whose rate of expansion would gradually slow down over time. Given sufficient time—actually, an infinite period of time—expansion should come to a halt. This condition is the flat universe predicted by inflation theory (compared to an infinitely dense or completely empty universe). The 1997–98 findings provided little support for the existence of a flat universe.

Those findings were based on the study of a group of supernovae known as Type Ia supernovae. Such supernovae are believed to be the final stage of white dwarf stars that attract matter at prodigious rates from a companion star and eventually blow themselves up in the manner of a gigantic hydrogen bomb. The Hubble telescope provided data on a number of such supernovae, and astrophysicists were able to compare these data with data from similar supernovae much closer to the Earth.

These observations made it possible, then, to compare the properties of supernovae that formed very early in the history of the universe (those observed by the Hubble) with those formed only very recently. The assumption is that all forms of the supernovae are fundamentally similar.

The most important finding in this research was that the rate at which young Type Ia supernovae are accelerating outward is significantly greater than the rate at which older supernovae of the same class are receding. In other words, the universe is expanding more rapidly now than it was many billions of years ago when it was much younger.

How can these results be explained? It is already clear that there seems to be far less matter in space than is needed to slow the expansion of the universe to the rate predicted by inflation theory. In the absence of any evidence for dark matter, astrophysicists have begun to ask whether it is possible that an amount of energy mathematically equivalent to the missing matter may be present in the universe. That energy could be equivalent to the missing matter because of the well-known relationship

These Hubble Space Telescope images pinpoint three distant supernovae, which exploded and died billions of years ago. *Courtesy of P. Garnavich (Harvard-Smithsonian Center for Astrophysics) and NASA.*

of mass and energy ($E = mc^2$). But in this case, the energy might be an entirely new and unusual form of energy—antigravity.

Antigravity

Antigravity is a force that has exactly the opposite effect on matter as gravity. That is, it tends to drive matter apart. It has been hypothesized in the past by various theorists. Perhaps the best-known suggestion for antigravity was Albert Einstein's theory of a cosmological constant, or *lambda*. Einstein invented the concept of a cosmological constant when he found that his General Theory of Relativity led to the prediction of a universe that continued to expand forever. He was uncomfortable with such a concept, and he invented the lambda constant in order to avoid such a conclusion. When Edwin Hubble discovered in 1929 that the universe was, in fact, expanding, Einstein was elated. He threw out the cosmological constant he invented, calling it "the biggest blunder of my life."

Inflation theory changed the rules of the game. It provided that the universe would expand, but that the rate of expansion would constantly decrease and approach zero as a limit. In order for this result to make sense, there would have to be enough matter in the universe to slow expansion (the "dark matter" problem), or some other explanation would have to be developed to explain why the universe's expansion had *not* slowed down (the cosmological constant).

It may be too soon to adequately assess the evidence from Type Ia supernovae. Critics had suggested that dust in the universe may have contaminated the more distant supernovae results. Or, that older Type Ia supernovae may not, in fact, be as similar to young ones as researchers have supposed. Members of two research teams studying the Type Ia supernovae have claimed, however, to have evaluated every possible error in their findings. In their most recent report, they claim that their results are valid with a statistical confidence of 98.7–99.99 percent.

If these findings are confirmed, astrophysicists will be confronted with a difficult dilemma. They will have to consider the possibility that an entirely new kind of force exists in the universe, an antigravitational force unlike anything now known. Or, they may have to consider modifications to inflation theory to account for the unexpected rates of expansion that these and other results now seem to suggest. Or, they may have to find other ways of dealing with the unexpectedly low density of mass and rates of expansion that observations appear to have confirmed.

References

Chang, Kenneth, "The Universe Flies Apart," <http://abcnews.go.com/sections/science/DailyNews/antigravity0226.html> accessed 30 May 1999.

Cowen, R., "The Cosmos' Fate: World without End," *Science News*, 3 January 1998: 4.

Glanz, James, "Astronomers See a Cosmic Antigravity Force at Work," *Science*, 27 February 1998: 1298–99.

Glanz, James, "Exploding Stars Point to a Universal Repulsive Force," *Science*, 30 January 1998: 651–52.

Glanz, James, "New Light on Fate of the Universe," *Science*, 31 October 1997: 799–800.

Gribbin, John, "Cosmology for Beginners," <http://epunix.biols.susx.ac.uk/home/John_Gribbin/cosmo.htm> accessed 30 May 1999.

"Introduction to Inflation" <http://map.gfsc.nasa.gov/html/inflation.html> accessed 30 May 1999.

Lemonick, Michael D., and J. Madeleine Nash, "Unraveling Universe," *Time*, 6 March 1995; also <http://cgi.pathfinder.com/time/magazine/archive/1995/950306/950306.cover.html> accessed 30 April 1999.

Musser, George, "Inflation Is Dead; Long Live Inflation," *Scientific America*, July 1998: 19–20.

Peterson, Ivars, "Circles in the Sky," and Ron Cowen, "Cosmologists in Flatland," (two-part series), *Science News*, 21 and 28 February 1998: 123–25 and 139–41.

Watson, Andrew, "Inflation Confronts an Open Universe," *Science*, 6 March 1998: 1455.

Antimatter

Sometimes the universe seems like a gigantic sock drawer. Astrophysicists always seem to be looking for missing objects, like the other half of a pair of socks. Dark matter is certainly one of the most important examples of something almost everyone believes in, but no one can find. Antimatter is another.

Antimatter is matter that resembles ordinary matter in every respect but one: electric charge. For example, a positron is exactly identical to an electron except for its charge (positive). Similarly, an antiproton is just like a proton except that it carries a negative charge rather than a positive charge. Antimatter, then, is made of positively charged electrons and negatively charged protons.

Antimatter may seem peculiar to those who were brought up knowing only about "ordinary" atoms made of positively charged protons and negatively charged electrons. But theories of cosmology clearly indicate that there is no particular reason for the universe to favor one kind of matter over the other. In fact, calculations show that the Big Bang should have resulted in the formation of exactly (or nearly exactly) equal amounts of matter and antimatter. If that's the case, why is it that scientists cannot now detect antimatter in the universe?

A partial answer for that question is based on objects relatively close to our own Solar System. Cosmological calculations show that there may have been a very slight bias in favor of matter over antimatter shortly after the Big Bang. Predictions are that for every 30 billion particles of antimatter, 30 billion plus one particles of matter were formed. In that case, the two forms of matter would have corresponded to each other exactly—except for one extra particle of matter.

Of those 60 billion original particles, none would have remained very long. When a particle collides with an antiparticle, the two disintegrate with the release of a large amount of energy. That energy is carried away for the most part by gamma rays. All that would have remained after the first few moments of the infant universe, then, would have been the excess particles of matter; not many on one scale, perhaps, but enough to generate the universe as we know it.

Searching for antimatter in regions relatively near the Sun is not too difficult. One only need look for characteristic emissions of gamma rays that would have been produced by the collision between matter and antimatter. No such evidence has ever been found, and physicists have been convinced for some time that antimatter does not exist in the cosmic neighborhood around Earth.

But what about the situation at more distant regions? Answering this question is a good deal more difficult. Matter-antimatter collisions that occurred early in the life of the universe would now be visible at only the most remote regions of the universe, but light reaching us from such collisions would have changed dramatically because of smearing by clouds of gas and dust and by the expansion of the universe itself. Thus, it is difficult to look at the most distant reaches of the universe and say whether they contain gamma ray remnants of early matter-antimatter annihilations.

For some time, however, astrophysicists have believed that such evidence might exist. The universe is filled with a diffuse gamma ray background for which there is currently no explanation. Perhaps that gamma ray background is the annihilation remnant for which scientists have been looking.

In 1997, a group of physicists announced that this scenario is not a realistic possibility. Here's how they reached their conclusion: One can assume that matter-antimatter annihilation occurred in galaxies or galactic clusters early in the history of the universe. It is possible to calculate the amount of energy released during such processes, as well as changes in that energy due to obscuring matter, inflation, and other effects. Ultimately, it is possible to estimate the amount of gamma ray energy that *should* be observable today if matter-antimatter annihilations did occur in the early universe. Such calculations are enormously laborious, but not conceptually difficult. As one researcher observed, "It's just a case of laboriously applying our knowledge to a very complicated thermody-

namic calculation" (Taubes, 226). In any case, the results of these calculations were that the observable gamma-ray background is far too weak to
have been caused by matter-antimatter annihilation. Such events would
have produced gamma-ray energies today that are at least five times as
great as those observed.

At the Kennedy Space Center, technicians observe the alpha-magnetic spectrometer (AMS-1)—an antimatter detector. *Courtesy of NASA.*

These findings do not absolutely rule out the possibility that antimatter exists elsewhere in the universe. From an observational standpoint, antimatter galaxies would look exactly like matter galaxies to an Earthbound observer. It is still possible that clumps of antimatter do exist somewhere in the universe where they do not come into contact with clumps of matter. We could be seeing both kinds of matter, isolated from each other, with no way of knowing that the two kinds of clumps differ from each other in their constitution.

Some researchers are still working to locate antimatter in the universe. Nobel Prize winner Samuel Ting, for example, has designed an antimatter detector scheduled to be placed on the International Space Station in the early 2000s.

The First Antiatom

While astrophysicists were puzzling over the absence of antimatter in the universe, particle physicists were taking the first steps in actually producing antimatter. In 1995, scientists at the Centre Européen pour la Recherche Nucléaire (CERN) Low Energy Antiproton Ring (LEAR) announced that they had produced the first antiatom. The antiatom was an atom of hydrogen, consisting of a single positron (antielectron) orbiting a single antiproton.

The antiatom was produced by directing a beam of antiprotons at a target of xenon atoms. On very rare occasions when the flood of antiprotons passed through a xenon atom, some of the antiprotons' energy was given up in the formation of an electron-positron pair. That kind of event is well known and is called *pair production*. On a few of the occasions when pair production occurred, the positron formed joined with one of the antiprotons to form an antiproton-positron pair: an antiatom of hydrogen.

The antiatoms formed were very unstable and collided with normal matter almost immediately. They had a lifetime of about 40×10^{-9} second, after which they were annihilated, with the release of gamma rays. The annihilation pattern captured on film provided the positive evidence that researchers needed to confirm that the antiatoms had actually been formed.

In their initial studies, the CERN researchers announced the formation of nine antiatoms as the result of billions of passes of the antiproton beam through the xenon target. Their next objective was to produce a larger number of antiatoms to see if the properties of antihydrogen are similar to those of normal hydrogen.

References

"Cosmic Antimatter Detection" <http://www-tele.fnal.gov/directorate/public_affairs/ams/ams.html> accessed 30 April 1999.

"First Atoms of Antimatter Produced at CERN" <http://www.cern.ch/Press/Releases96/PR01.96EAntiHydrogen.html> accessed 30 April 1999.

Tarlé, Gregory, and Simon P. Swordy, "Cosmic Antimatter," *Scientific American*, April 1998: 36–41.

Taubes, Gary, "Theorists Nix Distant Antimatter Galaxies," *Science*, 10 October 1997: 226.

Curved Space-Time

Albert Einstein first announced his General Theory of Relativity in 1905. For nearly a century, that theory has been providing physicists with hypotheses about the universe that would once have seemed ludicrous. Yet, hardly a year now goes by without some new evidence being reported that confirms some aspect of Einstein's "fantastic" world view.

Such was the case in the period 1997–98 when a series of reports confirmed Einstein's prediction of the relationship among mass, energy, and space-time. In Einstein's world view, the universe does not consist of some fixed volume of space, like a piece of plywood. Instead, he envisioned space (along with its fourth component, time) as a flexible entity, better represented by a sheet of rubber or flexible plastic. The shape of the space-time continuum could be affected, Einstein said, by the presence of energy or mass which, themselves, were influenced by the space-time continuum in which they existed and through which they moved.

As the 1990s drew to a close, reports appeared that confirmed this somewhat anti-intuitive conception of space. In one of these reports, scientists at the Massachusetts Institute of Technology (MIT) described their studies of X rays emitted from five regions of space believed to contain black holes. A black hole consists of the remnant of a star that has burned itself out and collapsed to a volume of vanishingly small dimensions. The object is said to be a "black" hole because the gravitational force it exerts is so great that not even light can escape from it.

According to Einstein's theory, all spinning objects have a tendency to drag the space-time continuum in which they are situated around with them. One could compare this effect with that observed by placing a spinning top into a pool of water. As the top continues to spin, the water around it is dragged along in a vortex with the top at its center.

For most spinning objects, the Einsteinian "drag" effect is too small to measure. But black holes are among the most massive objects known. Should the "drag" effect exist, black holes are the perfect place to look for it.

The approach used by the MIT team was to study X rays emitted in the vicinity of black holes. X rays are emitted when gases orbiting a black hole are pulled into it. The "drag" effect should be observable when the plane of the gases orbiting the black hole is inclined to the plane of the hole's rotation. In such a case, theory predicts that the pattern in which X rays are released from the black hole is altered. What an observer should see is a pattern of rapid oscillations in the X rays emitted from the region of the black hole. This pattern is what the MIT team believes it saw.

The problem with the MIT research is that it contains a relatively high degree of uncertainty. Indeed, most observers believe that the MIT results should be regarded as "suggestive" and that additional confirmatory studies will be necessary before they can be completely accepted.

At nearly the same time the MIT results were announced, however, similar findings were reported by a team from the Astronomical Observatory and the University of Rome. This team chose to study neutron stars rather than black holes. The argument for neutron stars is very similar to that for black holes. Neutron stars are very dense objects formed when certain types of stars burn out and "die." Like black holes, they produce enormous gravitational fields and should cause the space-time continuum to warp as they rotate on their axes.

The Rome team found a pattern of X-ray emissions similar to that reported by the MIT team. That is, the Rome researchers found a pattern of very rapid oscillations in the pattern of X rays emitted by rotating neutron stars. They attributed this effect to the neutron stars' dragging the space around them as they rotated on their own axes. Like the MIT results, however, the Rome findings included a relatively high degree of uncertainty.

In early 1998, a team of Italian, Spanish, and Greek-American scientists reported on yet another example of space-time drag caused by rotating bodies. In this case, the rotating body was much closer at hand: the Earth.

This team argued that the rotation of the Earth on its axis should drag space-time in its vicinity along with it. The effect would be very much smaller than for black holes or neutron stars, of course, but the effect would be much easier to measure.

The objects used to measure drag in this case were two satellites originally launched to study the size and shape of the Earth, LAGEOS I and LAGEOS II (LAGEOS is an acronym for Laser Geodynamics Satellite). Researchers began their studies with a very precise knowledge of the shape of the Earth's gravitational field, developed by a consortium of researchers over the previous four years. They then aimed laser pulses from the Earth to the LAGEOS satellites to determine their exact positions. The technique they used allowed them to find those positions with an error of less than a centimeter.

When all known effects on the satellite orbits had been accounted for, researchers found that those orbits still contained a shift of about two meters, an effect they attributed to the drag caused by the Earth's spin. In this case, the uncertainty expressed by this measurement is about 20 percent. The team is continuing this line of research and hopes to reduce measurement errors and uncertainty even more.

References

Ariza, Luis Miguel, "Einstein's Drag," *Scientific American*, July 1998: 24–25.

Ciufolini, Ignazio, "Test of General Relativity and Measurement of the Lense-Thirring Effect with Two Earth Satellites," *Science*, 27 March 1998: 2100–03.

Cowen, R., "Einstein's General Relativity: It's a Drag," *Science News*, 15 November 1997: 308.

Glanz, James, "X-rays Hint at Space Pirouette," *Science*, 7 November 1997: 1012–13.

Mukerjee, Madhusree, "Girth of a Star," *Scientific American*, November 1997: 22.

Peterson, I., "X-ray Flashes Illuminate General Relativity," *Science News*, 25 April 1998: 261.

Solar Physics

Astronomers have been accumulating data about the nearest star, our Sun, for hundreds of years. Now it is the turn of solar physicists to try making sense out of much of those data. Invaluable in this effort has been the availability of space probes and satellites that provide an even closer and more detailed look at solar structures and processes.

One example of the tools solar physicists now have at their command is the Solar and Heliospheric Observatory (SOHO), whose operation is dedicated to an around-the-clock and around-the-calendar study of the Sun. Another example is the space probe Galileo, which has had a number of tasks, one of which was a study of the Sun as it made a pass around the star on its way to Jupiter.

Over the past few years, these tools have revealed new information on the way the Sun behaves. One example is the discovery of the mechanism by which the solar wind is generated. *Solar wind* is a term used to describe the flow of particles constantly emitted from the Sun's surface. That flow is emitted continuously and spreads out over much of the Solar System.

The solar wind actually consists of two components, a "fast" wind that appears to emanate from the solar poles and a "slow" wind that appears to originate close to the solar equator. No one has been able to explain, however, what the source of the solar wind is.

In late 1997, solar physicists reported new data that answered at least part of the puzzle regarding the solar wind, the part involving the slow wind. It appears that the slow wind is produced from the Sun's corona, the very hot halo of gases that surrounds the Sun at a distance of about 5,000 kilometers above the Sun's surface. Instruments aboard SOHO detected particles being ejected from tall, narrow structures called *stalks* located at the top of magnetic loops. These magnetic loops originate on the Sun's surface and extend far into the corona. Surveys of the location of stalks and the outward flow of particles that make up the solar wind showed the two regions to be nearly identical. The evidence was strong enough to convince many solar physicists that the source of the slow solar wind had been found.

Data on the fast wind proved more difficult to interpret, however. Solar physicists have generally assumed that the fast wind was produced at the solar poles. In these regions, the magnetic loops that reach from the Sun's surface to the corona are less well defined. Particles escaping from the Sun are less constrained by the magnetic fields and can escape with greater velocity.

At least that was the theory. The SOHO and Galileo findings described above, however, raised doubts about this explanation. It appears that the fast wind is being produced not just at the poles, but all over the Sun's surface. If such is the case, there may be "holes" in the surface where magnetic fields are not strong enough to slow down the movement of escaping particles. This interpretation of the fast wind creates great concern among solar physicists because it involves rejecting a long-held notion about the location of the fast wind and requires the creation of a new theory for how fast-moving particles can escape from the Sun's surface.

The problem of the solar wind is closely associated with a second difficulty that may also have been solved by the SOHO experiments. That problem is the temperature of the corona (the outermost region of the Sun's atmosphere). Astronomers have known for a long time that the temperature of the corona is very high, several millions of degrees. By comparison, the temperature of the solar surface is no more than a few thousand degrees. The question, then, is how the corona can become so hot.

The most obvious answer, of course, is that heat flows from the Sun to its outermost regions, raising the corona's temperature. That explanation cannot be correct, however, since physicists know that heat cannot flow from a cooler body (the solar surface) to a warmer body (the corona).

Observations made by SOHO now seem to provide an answer to this dilemma. It appears that the Sun's surface is covered with thousands of individual magnetic fields. One researcher has referred to this structure as a "magnetic carpet" that covers the face of the Sun.

Each magnetic field in that magnetic carpet consists of two poles, a north pole and a south pole, connected by lines of force that extend upward into the corona. Particles ejected from the surface of the Sun spiral upward along the magnetic lines of force into the corona. There they release the energy they picked up in the Sun's interior before being released in the form of the slow wind or being returned to the solar surface along the lines of force.

The data for this analysis came from long-term studies conducted from SOHO. In a sense, these studies were comparable to very sophisticated "home movies" that watched documented changes taking place on the Sun's surface over long periods of time.

These studies showed that the magnetic fields of which the so-called magnetic carpet is made are highly transient. They are born, grow to maturity, and disappear in a few tens of hours. While providing an answer to the question of the corona's temperature, these findings also raise new questions. For example, physicists would like to know what physical mechanisms are taking place within the Sun's interior to make possible this rapid creation and destruction of magnetic fields.

References

Cowen, R., "SOHO Craft Helps Solve a Solar Mystery," *Science News*, 8 November 1997: 295.

Day, Charles, "SOHO Observations Implicate 'Magnetic Carpet' as Source of Coronal Heating in Quiet Sun," *Physics Today*, March 1998: 19–21.

"ESA: Solar Mystery Nears Solution with Data from SOHO Spacecraft," M2 PressWIRE, 17 November 1997; available on the Electronic Library, <http://www.elibrary.com> accessed 30 April 1999.

Glanz, James, "Two Spacecraft Track the Solar Wind to Its Source," *Science*, 17 October 1997: 387–88.

"NASA Space Science News" <http://www.astronomynews.com> accessed 30 April 1999.

"SOHO News" <http://umbra.gsfc.nasa.gov/sdac.html> accessed 30 April 1999.

Further Reading

Bernstein, Jeremy, *An Introduction to Cosmology*, New York: Prentice Hall, 1997.

Cornell, James, ed., *Bubbles, Voids, and Bumps in Time: The New Cosmology*, Cambridge: Cambridge University Press, 1992.

Foukal, Peter, *Solar Astrophysics*, New York: John Wiley and Sons, 1990.

Hawkins, Michael, *Hunting Down the Universe: The Missing Mass, Primordial Black Holes, and Other Dark Matters*, Reading, MA: Perseus Books, 1999.

Hogan, Craig, *The Little Book of the Big Bang: A Cosmic Primer*, New York: Copernicus Books, 1998.

Kragh, Helge, *Cosmology and Controversy: The Historical Development of the Two Theories of the Universe*, Princeton, NJ: Princeton University Press, 1999.

Padmanabhan, T., *After the First Three Minutes: The Story of Our Universe*, Cambridge: Cambridge University Press, 1998.

Rees, Martin J., *Just Six Numbers*, New York: Basic Books, 1999.

Silk, Joseph, *A Short History of the Universe*, New York: W. H. Freeman, 1994.

Wheeler, John Archibald, *A Journey into Gravity and Spacetime*, New York: W. H. Freeman, 1999.

BOSE-EINSTEIN CONDENSATES

High school science students learn that matter can exist in any one of four forms: solid, liquid, gas, or plasma. These forms of matter occur commonly in the natural world, although the first three are the most familiar on Earth. Plasmas are found only under relatively unusual conditions, such as at the center of a fusion (hydrogen) bomb or at the center of a star.

Now scientists have discovered a fifth form of matter that occurs nowhere in the natural world except in the most sophisticated physical laboratories on Earth. That fifth form of matter is known as a Bose-Einstein condensate.

Bose-Einstein condensates (BECs) were predicted in the 1920s by the Indian physicist Satyendra Nath Bose and the Austrian-American physicist Albert Einstein. Expanding on calculations originally made by Bose, Einstein predicted that atoms cooled to very low temperatures would behave in a manner impossible to observe on the macroscopic scale. He said that those atoms would all take on identical quantum mechanical characteristics such that it would be impossible for an observer to tell them apart. A collection of such atoms would, in effect, behave like a single very large atom, many times the size of the individual atoms of which it is composed.

Quantum Properties of Matter

The Bose-Einstein prediction was based on a fundamental dichotomy that has existed in physics for nearly a century. The basis of that dichotomy is that humans can normally observe matter only at a macroscopic scale: a few grams, milligrams, micrograms, or nanograms of matter at a time. But even the smallest sample of matter consists of millions and billions of atoms. Each of these atoms behaves as an individual unit, like an individual bee in a swarm or hive.

But physicists have long known that the behavior of individual atoms can be explained correctly only by using quantum mechanics, a theory that is essentially useless on the macroscopic level. One fundamental principle of quantum mechanics, for example, is the Heisenberg Uncertainty Principle. The Uncertainty Principle says that it is never possible to measure both the position and velocity of a particle with perfect accuracy. The more accurately one knows one of these properties, the less accurately the other is known. This principle is an absurdity at macroscopic levels, of course, where physicists daily measure with high precision both the position and velocity of objects.

But BECs also provided scientists with a mechanism, at least in theory, by which the concepts of quantum mechanics could be observed and tested at macroscopic levels. If a batch of atoms could be cooled sufficiently, they could all be made to assume a common set of quantum properties. Methods could then be developed—again, in theory—to observe those properties, just as the macroscopic properties of matter are observed.

The fundamental problem facing physicists in the 1920s was that the temperatures at which BECs formed were incredibly low. Einstein's calculations, for example, showed that such condensates would form at only a few millionths of a degree above absolute zero (the temperature at which all atomic and molecular motion comes to an end).

The Technology of BEC Formation

For nearly 70 years, then, the possibility of creating a BEC was an intriguing theoretical speculation, but nothing more. Then, in the 1970s, physicists began to develop techniques for cooling atoms to temperatures far lower than had ever been imagined. By the early 1990s, those techniques had been highly developed. The 1997 Nobel Prize in Physics was awarded to three physicists—Steven Chu, Claude Cohen-Tannoudji, and William Phillips—who perfected these techniques.

The procedures invented by Chu, Cohen-Tannoudji, and Phillips made possible the formation of a BEC in a two-step process. In step one of that process, a small mass of atoms is placed in a container. Although any atom could be used in theory, physicists had found that alkali atoms were more likely than atoms of any other kind to form BECs. Early experiments on the formation of BECs, then, used rubidium and sodium atoms.

The sample of atoms within the container has, of course, a variety of quantum mechanical characteristics. Any measurements made on that sample of atoms will detect the summed average of all those characteristics. It will be impossible to know the quantum mechanical properties of any single atom within the mixture.

The first step in cooling these atoms is to bombard them with laser beams arranged around the outside of the container. The way in which laser beams cool atoms is as follows: A laser beam consists of photons that, because of their motion, have a certain amount of momentum.

When a photon strikes an atom, it transfers some of that momentum to the atom. If the photon and atom are traveling in the same direction, the collision increases the speed of the atom. If the photon collides with the atom head-on, it decreases (slows down) the speed of the atom.

One of the fundamental developments made by Chu, Cohen-Tannoudji, and Phillips was to find a way of arranging laser beams so that *every* atom was likely to be slowed down by an interaction with photons. They set up one laser on each of the six outside faces of a cubic box. Each laser was aimed inward at the atoms inside the box. No matter which direction an atom was traveling, it collided with a laser beam from one of the six faces of the box.

In addition, each laser was "tuned" to the frequency of the atoms in the sample. That is, the frequency at which an atom was traveling was exactly matched by the frequency of the laser beam. In this way, the atom absorbed all of the energy carried by the laser beam, causing it to slow down as efficiently as possible.

Thus, the longer the laser beams were shined on the sample within the container, the less rapidly atoms within the sample moved. As the atoms slowed down, so did their kinetic energy and their temperature. By the use of laser cooling alone, a sample of atoms could be cooled to a temperature of about 40-millionths of a degree above absolute zero. This temperature is extraordinarily low and a remarkable achievement in and of itself—but it is still not low enough for the production of a BEC.

A second stage of the cooling process, therefore, is needed. In this stage, additional cooling is produced by magnetic confinement and evaporative cooling. Magnetic confinement is essentially a system for holding the very cold atoms in place after the laser beam has been turned off. The process depends on the fact that all atoms spin on their own axes, producing a weak magnetic field around themselves. The atoms are, in a sense, like very tiny bar magnets. If a magnetic field is imposed on a collection of atoms, it will hold those atoms suspended in place.

The final step, then, is evaporative cooling, a process well known from everyday life. A container of water left open to the air, for example, gradually becomes cooler as the water evaporates. Cooling occurs because the particles within the container with the greatest amount of energy are able to escape (evaporate) first. As they evaporate, particles of lower energy (less velocity) are left behind.

Within the liquid, particles exchange energy with each other by collisions. When two moving water molecules collide with each other, for example, one particle gains energy and the other loses energy in the collision. The more energetic (faster-moving) particle is more likely to escape by evaporation, while the less energetic (slower-moving) particle is left behind.

In the experimental set-up described above, atoms confined by a magnetic field undergo a similar process. Those atoms with slightly more energy collide with slower-moving particles, absorb some of their energy, and eventually escape from the magnetic trap. Over a period of time, the energy of the particles remaining within the trap becomes less and less. In the best magnetic traps invented, evaporative cooling reduces the temperature of the sample of atoms to about 100-billionths of a degree above absolute zero, or less. At this temperature, the formation of a BEC is possible.

Observation of the First BEC

The first successful production of a BEC using this technology was announced in 1995 by researchers at the Joint Institute for Laboratory Astrophysics (JILA) in Boulder, Colorado, a joint program of the University of Colorado and the National Institute of Standards and Technology. The research team was directed by Eric Cornell and Carl Wieman. The JILA team produced a BEC that consisted of about 2,000 atoms of rubidium at a temperature of 20×10^{-9} degrees kelvin.

The team was actually able to take a photograph of the rubidium condensate that had formed. Computer analysis of this photograph showed that a significant number of rubidium atoms at the center of the magnetic trap had been cooled to a point at which they were identical to each other in terms of their quantum mechanical characteristics. The presence of these atoms is represented by the tall, narrow peak at the center of the computer plot.

This plot depicts a particularly interesting point about the JILA accomplishment. According to Bose-Einstein theory, all of the atoms trapped within the magnetic field should have exactly identical quantum mechanical properties—they should be entirely indistinguishable from each other. A computer plot of such a sample should show a very narrow line representing all of the atoms in the sample. The fact that the actual computer plot is not precisely a line is a manifestation of the Uncertainty Principle. Had a line been produced on the computer plot, one could conclude that the position and velocity of every atom in the sample was known with perfect precision. The Uncertainty Principle does not permit such a measurement. The computer plot is, therefore, a rather remarkable macroscopic depiction of a key quantum mechanical principle.

It is this aspect of the Cornell-Wieman experiment (and others similar to it) that is of such great interest to physicists. The ability to produce condensates of measurable size provides an opportunity to study quan-

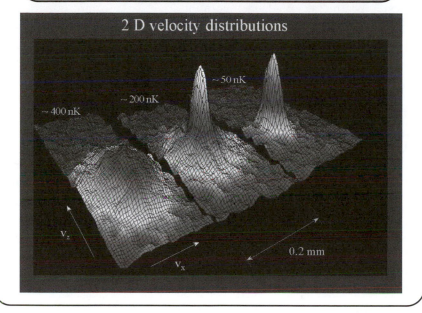

A computer-processed image showing the formation of a Bose-Einstein condensate. *Courtesy of JILA/Michael Matthews.*

tum mechanical phenomena on a macroscopic scale, an opportunity that seemed impossible only a decade ago.

The Cornell-Wieman discovery may have practical applications, as well. For example, the collection of rubidium atoms the JILA team produced is similar in some respects to the collection of photons produced in a laser. In a laser, all photons have precisely the same quantum mechanical properties, such as direction, frequency, and velocity. The ability to collect and focus such a beam of photons is what makes lasers so valuable in so many different ways in the real world. The JILA accomplishment means that it may be possible, also, to develop a "laser" made of atoms with properties similar to those of light lasers. Progress along these lines is reported below.

The Cornell-Wieman breakthrough has been followed by similar accomplishments using other alkali elements, such as lithium and sodium. Perhaps the most exciting of these follow-up successes was that of a research team at the Massachusetts Institute of Technology under the direction of Thomas J. Greytak and Daniel Kleppner. Greytak and Kleppner had been working for 20 years to produce a Bose-Einstein condensate using hydrogen. They thought that hydrogen, the simplest of all elements, might be the first to yield to the techniques needed for capturing and holding a group of atoms in a single quantum state.

They turned out to be incorrect in this prediction, but their 1998 success in producing a hydrogen condensate was important nonetheless. In the first place, the condensate they produced was the largest yet made, consisting of more than 100 million atoms. In addition, the ability to form a hydrogen condensate was particularly significant because more is known about hydrogen than any other element, and the condensate form of the element is expected to yield even more fundamental information about the nature of matter.

References

Anderson, M. H., et al., "Observation of Bose-Einstein Condensation in a Dilute Atomic Vapor," *Science*, 14 July 1995: 198–201.

"Bose-Einstein Condensation" <http://www.colorado.edu/physics/2000/bec> accessed 30 April 1999.

Cornell, Eric A., and Carl E. Wieman, "The Bose-Einstein Condensate," *Scientific American*, March 1998: 40–45.

Goss Levi, Barbara, "Work on Atom Trapping and Cooling Gets a Warm Reception in Stockholm," *Physics Today*, December 1997: 17–19.

Kestenbaum, David, "Hydrogen Coaxed into Quantum Condensate," *Science*, 17 July 1998: 321.

"Physicists Create New State of Matter at Record Low Temperature" <http://jilawww.colorado.edu/www/press/bose-ein.html> accessed 30 April 1999.

"The Ultimate Chill-Out Proves Einstein Right Once Again" <http://whyfiles.news.wisc.edu/052einstein/bose.html> accessed 30 April 1999.

Weiss, P., "Hydrogen Atoms Chill to Quantum Sameness," *Science News*, 25 July 1998: 54.

The Atomic Laser

It is not unusual for physicists to make important discoveries about the nature of matter and energy that have few obvious practical applications or none at all. The discovery of the laser in 1960 is an example. Theodore Maiman's discovery that beams of intense, coherent, visible light could be produced was an enormously fascinating breakthrough, but few physicists could have predicted the dozens of applications that discovery would have in the 1990s.

The same statement can probably be made for the discovery of Bose-Einstein condensates. This discovery has certainly been one of the major breakthroughs in the last few decades, but how many physicists can predict today what applications this discovery will have in a generation.

Interestingly enough, one such application has already appeared: the atomic laser. An atomic laser is a device that produces a beam of atoms with identical quantum properties; it is somewhat similar to the beam of photons with identical quantum properties produced in an optical laser. The first announcement of a successful atomic laser came in 1997 from a team of researchers at the Massachusetts Institute of Technology (MIT) under the direction of Wolfgang Ketterle.

Principles of an Optical Laser

In order to understand the operation of an atomic laser, it is helpful to review the principles of an optical laser. An optical laser consists of two primary components: a medium in which photons are generated and a reflecting system that transmits those photons back and forth through the medium. When an external source of energy (such as an electrical current) is provided to the medium, a certain fraction of electrons in the ground state of atoms in the medium are elevated to an excited state. After spending a brief moment in that excited state, the electrons return to the ground state, each releasing a photon of energy in the process.

The laser system is designed so that the photons produced are then reflected off one of the mirrors placed at either end of the device. The reflected photons act as a new source of energy, promoting other electrons in the ground state of medium atoms to their excited state. When these electrons again return to their ground state, they also release photons.

In this sequence of events, more and more photons are produced so that the original weak pulse of radiant energy is amplified. This amplification occurs because of the stimulated emission of photons at some later stage by photons created at some earlier stage. Thus does the laser derive its name; it produces *l*ight *a*mplified by *s*timulated *e*mission of *r*adiation.

Principles of an Atomic Laser

Atomic lasers are possible because of the dual nature of matter and energy; that is, energy behaves at times as if it consists of tiny particles. Photons are an example of these "particles of energy." At the same time, matter sometimes behaves as if it is a form of energy. When electrons travel through space, for example, they travel as waves that can be described by their frequency, wavelength, and amplitude, just as light waves are described. It is possible to make electron waves interfere and diffract in just the way one can make light waves interfere and refract.

One of the conditions necessary in creating an optical laser is coherence. In such lasers, all of the photons produced are exactly the same because they are all produced from identical atoms in identical ways. As they travel through and out of the laser tube, they travel in waves that overlap—or are coherent with—each other.

In order to make an atomic laser, it also is necessary to produce waves that are coherent with each other. That is, all of the atoms that make up such a wave must have exactly the same quantum mechanical properties. Such a condition would have been difficult—if not impossible—to meet without the discovery of Bose-Einstein condensates. One of the properties of the atoms in such a condensate, of course, is that they all have identical quantum mechanical properties. The technical problem, then, is to find a way to extract the bundle of atoms that makes up a BEC within a laser-magnetic trap and transmit it through space as a wave.

Atomic Lasers versus Optical Lasers

Some fundamental differences between atomic lasers and optical lasers exist, of course. For one thing, the number of photons that can be generated in an optical laser is unlimited. Each time an electron is elevated to an excited level, it will produce another photon when it returns to its ground level.

But atoms are a form of matter. It is not possible to create atoms in the same way that photons are created. In such a case, how is it possible to obtain amplification, one of the key elements of a laser?

The MIT team found a method for solving this problem. When a bunch of atoms is trapped as a Bose-Einstein condensate, they all have, by definition, identical quantum mechanical properties. What the MIT team did was to apply an external field of radiofrequency (rf) radiation to a group of atoms trapped in a BEC. The rf field caused some of the atoms in the BEC to reverse their spin. They were no longer trapped by the external magnetic field, and they dropped out of the BEC.

The term "dropped out" is appropriate because the atoms released from the trap moved in the direction dictated by gravity. As with any other source of matter, they responded to the force of the Earth's gravitational field. And this fact points out another of the differences between atomic and optical lasers. The beam from an optical laser can, of course, be pointed in any direction whatsoever, just as one can shine a flashlight in any direction. But the beam of an atomic laser can be directed only downward, in the direction of gravitational forces. MIT researchers are currently working on methods to alter this limitation of the atomic laser. They are exploring the possibility, for example, of using magnetic fields as "mirrors" off of which an atomic beam can be reflected.

The other problem that faced developers of the atomic laser was that of amplification. As noted above, it is not possible to create a larger number of atoms than originally existed in the BEC, the way one can create larger numbers of photons in an optical laser. Yet, when the MIT team first tested its atomic laser, it made an interesting observation. After the rf field was turned on, a very weak atomic beam was emitted from the BEC. Over a period of time, however, the beam grew more and more intense. In other words, the performance of the atomic laser was similar to that of an optical laser.

Researchers are not certain how this amplification occurs. Their current theory is that atoms that have been released from the BEC "want" to return to the condensate because it represents a lower energy level, and all particles "want" to return to their lowest possible energy level. The only way a particle can return to the BEC, however, is to give off some energy. It does so by transferring energy to other atoms near to it that have also been released from the BEC. These atoms, then, gain energy and velocity. As this process is repeated time after time, the intensity of atoms released from the BEC continues to increase, resulting in an amplified, coherent beam of atoms that resembles a similar beam of photons produced in an optical laser.

Possible Applications

Predictions about the possible applications of an atomic laser are at about the same point that predictions for the optical laser were in 1960. One obvious application for the atomic laser would be to lay down thin layers of a material by means of the atomic beam from such a laser. This technology is already being explored as a way of working with very small structures that make up nanotechnology today.

At the moment, the most obvious applications of the atomic laser are in the field of basic research. The device should be useful, for example, in measuring some of the fundamental constants of nature with a precision never before possible.

But a number of problems must be solved before other applications of the atomic laser become apparent. For example, the MIT atomic laser produces only short bursts of atoms, in contrast to an optical laser, which produces a continuous beam of photons. Also, the problem of beam direction, discussed above, must be solved. All in all, it appears that few physicists are willing to make many predictions about the long-term applications of the atomic laser, although such applications will certainly occur.

References

Cooke, Robert, "Atomic Laser Created," *Newsday*, 27 January 1997: A8.

Hellemans, Alexander, "Atom Laser Shows That It Is Worthy of the Name," *Science*, 13 February 1998: 986–87.

"MIT Researchers Create Rudimentary Atom Laser" <http://www.aip.org/physnews/preview/1997/alaser/text.htm> accessed 30 April 1999.

Wilson, Jim, "Have Atoms, Will Travel," *Popular Mechanics*, July 1997: 36–37.

Further Reading

Griffin, A., D. W. Snoke, and S. Stringari, eds., *Bose-Einstein Condensation*, Cambridge: Cambridge University Press, 1996.

GEOPHYSICS

One might think that there is nothing new to learn about the Earth. After all, scientists have been studying our planet home for hundreds of years. But, of course, there is always something new to be learned about any aspect of nature. And the Earth is no different.

One of the inherent problems in studying the Earth is that observers are, in a sense, imbedded within their subject. It is difficult, for example, to think of the Earth as a spheroid when our senses tell us the planet is flat. It is hardly surprising, then, that some of the most interesting discoveries made about the planet recently have come from data supplied by space probes and satellites that have the ability to look back down on the Earth. These instruments provide scientists with a new platform for conducting terrestrial research that was not available until quite recently. Other instruments for analyzing the interior of the Earth and microscopic characteristics of its surface have also contributed to new knowledge.

Rates of Rotation

One area in which new information about the Earth has been acquired has to do with the planet's rotation rate. Over a period of many centuries, observers have been able to develop a more and more sophisticated understanding of the factors affecting this rate. That information is of some interest, at least partly because it is used in determining a fundamental time unit for humans: the day.

A day is defined as the time it takes for the Earth to make one complete rotation on its axis, compared to some standard. If the period of rotation is determined by the time it takes for the Sun to appear in exactly the same spot overhead, the unit of time is known as the *solar day*. If some fixed object in the skies, such as a specific star, is chosen, the time unit is called the *sidereal day*.

As technology has become more sophisticated, the measurement of the length of a day (LOD) has become more and more precise. Measurements have gone from hours (24 hours) to the exact number of hours, minutes, seconds, and fractions of a second.

Scientists now know that a number of factors affect LOD measurements; that is, the Earth's rate of rotation is not an absolute constant, but a variable that is affected by a number of factors. For example, the gravitational attraction between the Earth and the Moon causes the

Earth's rate of rotation to slow down and increase the LOD by about two seconds per century. Other factors thought to have an influence on the LOD include earthquakes, movement of groundwater, ocean currents, plate tectonics, and the shifting of matter in the Earth's interior.

For some time, scientists have known that the most important of these factors, by far, is the Earth's atmosphere. Air movements in the atmosphere cause friction with the Earth's surface and affect its rate of rotation. In general, the solid Earth and its atmosphere act as an integrated system such that the total momentum of both components is essentially a constant. That is, as the Earth itself loses momentum, the atmosphere increases its momentum by nearly the same amount.

The problem is that the balance between atmospheric momentum and planetary momentum is not exact. Other factors (such as those listed above) also influence the rate of rotation and, therefore, the LOD. In 1998, scientists at the Jet Propulsion Laboratory (JPL) at the California Institute of Technology reported findings that connected ocean movements with the Earth's rate of rotation.

The first step in this research was to confirm as precisely as possible the LOD. This measurement can now be made with very high precision by means of a technique known as very long baseline interferometry (VLBI). VLBI makes use of an array of radio telescopes to measure very precisely the moment at which a given point on the Earth passes under a reference point located outside of our galaxy.

The next step of the research is to look for factors that may affect the Earth's rate of rotation and estimate the factor that appears to have the greatest effect, after the atmosphere, on that rate. This estimate can be made by assuming that the vast majority of the Earth's angular momentum (call that variable M_E) is balanced by the angular momentum of the atmosphere (call that variable M_A). Then, it is known that:

$$M_E = M_A + M_1 + M_2 + \ldots M_n$$

with M_A making up, by far, the greatest part of M_E.

The JPL researchers considered a number of factors that might make up M_1, M_2, and so forth. Such factors would include the oceans, polar ice, and movement of the Earth's tectonic plates. A likely candidate for the next-most-important source of angular momentum is the Earth's oceans. Like the atmosphere, the water in the oceans slides back and forth over the Earth's surface, causing friction that could slow the Earth's rotation rate.

The JPL team did not have adequate data about the oceans to make the necessary determinations in this case. But the scientists were able to use models of the way that water circulates in the ocean that had been developed more than a decade earlier and had been increasingly refined ever since. When they determined the angular momentum of the oceans, estimated from these oceanic general circulation models (OGCMs), they found that they matched very closely the angular momentum missing from the general equation above. That is, it appears that the circulation of ocean water is the second most important factor, after the atmosphere, in affecting the rate at which the Earth rotates.

Earth Oscillations

The Earth's crust moves. Those who have witnessed an earthquake need not be reminded of that fact. Earthquakes occur when tectonic plates move. Tectonic plates are enormous masses of rock on which the continents rest. These plates are constantly colliding with each other, moving away from each other, and sliding back and forth against each other.

But the crust is moving in other ways also. It is vibrating up and down, back and forth, in movements known as *oscillations*. These oscillations are much more difficult to observe than are earthquakes. They tend to occur on a more modest scale.

Geophysicists have long wondered about the exact nature and cause of these oscillations. The most common view has been that they are the lingering remnants of earthquakes. They were first observed in 1960 after a 9.5 earthquake in Chile. The oscillations traveled around the Earth, like the ringing of a bell, with a period of about an hour, for days after the quake itself. Since the Chilean event, seismologists have observed similar global oscillations in connection with smaller earthquakes.

In 1998, Japanese researchers reported new findings on the nature of apparently continuous Earth oscillations of very small magnitude. Interestingly enough, these studies were initiated by an astronomer, Naoki Kobayashi, who had studied oscillations of the Sun and was curious about possible comparable effects on Mars and Venus. In pursuing this interest, he decided to see what he could learn about oscillations on our own planet Earth.

The approach taken by researchers in attacking this problem was to search for seismically "quiet" days in existing earthquake records. Quiet days are periods during which no earthquakes have been reported anywhere in the world. Such days are actually quite rare, and the Japanese team decided to use only 60 days out of eight years of records. If traditional arguments were true, no movement of the Earth's crust should be detectable during these periods.

A second team of Japanese researchers studied the same problem with a somewhat different approach. They studied records produced by a highly sensitive gravimeter (a device for measuring changes in the Earth's gravitational field) located at the South Pole. The gravimeter consisted of a niobium metal ball suspended in air above a metal plate by means of a magnetic field. This arrangement is able to detect very small changes in the Earth's gravitational field and, hence, in any oscillations that may occur in the Earth's crust.

Both researchers reported the same discovery: When all possible crustal movements attributable to known earthquakes were subtracted, a very small oscillation in the Earth's crust was still detectable. That background oscillation had a period ranging from about 2 to 8 minutes and an amplitude of about 10^{-8} meter, a dimension comparable to the diameter of a molecule.

These reports have not been accepted by all geophysicists. Some scientists think that the background oscillation detected by the Japanese teams is somehow connected with earthquakes—they are not convinced that any new phenomenon has been detected.

Those who accept the conclusions of this research have begun to ask what forces could account for the background oscillation. One suggestion has been that movements of the atmosphere—winds, in particular—may account for this effect. The Japanese researchers themselves point out that the Earth's surface is constantly being bombarded by downward movements of air. While these movements are not very powerful compared with the crustal rocks they strike, they may be sufficient to produce the very small reverberations observed in these studies.

Composition of the Earth's Core

Those who study the Earth's core share some of the same problems encountered by astronomers. First, the objects of such studies (stars and galaxies on the one hand, and the core on the other) are far too remote to be examined firsthand. Second, the conditions in which the objects being explored exist are extreme, and often difficult or impossible to reproduce in the laboratory. Yet, knowledge of both the farthest reaches of the universe and the interior of our own planet continues to develop because of ingenious techniques developed by physicists for their research.

In late 1997, a Japanese graduate student, Takuo Okuchi, offered a revolutionary new proposal for one of the "core" problems of interest to geophysicists—hypothesizing the composition of the Earth's core.

Most beginning science students learn two basic facts about the Earth's core: first, that it consists of two regions, an inner and outer core; and second, that it is made up largely of iron and nickel. Of course, geological scientists know a good deal more about these matters. For example, they have long known that the core is chemically more complex than suggested by the statement above. The density of the core is actually about 10 percent less than it should be if it were pure iron. In order to explain this observation, geological scientists have usually hypothesized the presence of one or more lighter elements in the core. The elements most commonly suggested have been hydrogen, carbon, oxygen, sulfur, and silicon.

Of these five elements, sulfur and silicon have been considered the most likely candidates. Hydrogen and carbon would probably be too volatile to have remained in the core during its formation, and oxygen does not dissolve in iron well enough to allow it to remain in the core. That leaves sulfur and silicon as the elements most likely to have combined with iron in some way so as to have allowed them to remain in the core.

Okuchi's revolutionary idea was that hydrogen, not sulfur or silicon, might be the light element that accounts for the core's lower-than-expected density. He based his suggestion on earlier findings that hydrogen will combine with iron at very high pressures to produce an intermetallic compound (a compound consisting of two metals) with the general formula FeH_x. The problem had been that such compounds are stable only at pressures of greater than 3.5 GPa (gigapascal, or billion pascal). At pressures less than this amount, the iron hydrides decompose to yield free iron and hydrogen.

As part of his doctoral studies, Okuchi designed a method for determining the chemical structure of FeH_x. He mixed iron, magnesium oxide (MgO), brucite ($Mg(OH)_2$), silica glass (SiO_2), silicic acid ($SiO_2 \cdot 4H_2O$), and liquid water in a small platinum capsule and then raised the pressure on this mixture to more than 5 GPa. Next, he "quenched" the mixture by releasing the pressure very quickly, at the rate of 1.5 GPa per second. Under these conditions, hydrogen gas formed by the decomposition of FeH_x had no chance to escape from the mixture but was, instead, trapped as bubbles within the solid iron that formed.

Finally, Okuchi determined the amount of trapped hydrogen gas and, from that value, calculated the empirical formula for the iron hydride that had been formed at higher pressures. He determined that the formula ranged from $FeH_{0.33}$ for the solid phase to $FeH_{0.4}$ for the liquid phase of the hydride.

This finding prompted Okuchi to postulate a way that hydrogen present during the formation of the Earth could have been incorporated into the core. Assuming that hydrogen was then present in the form of water, Okuchi calculated that 95 percent of all the water present when the planet was formed eventually became part of the core. In such a case, the vast majority of the hydrogen on the Earth today is to be found not in the hydrosphere (the oceans and other sources of water), but in its core.

As would be expected, Okuchi's results have not been accepted without question by other geophysicists. Some colleagues have raised doubts about his methods for analyzing the composition of the iron-hydrogen mixture produced at high pressures. Others doubt that conditions in the primordial Earth would have allowed the "soaking up" of water by an iron core, as Okuchi speculates. As with most issues in science, only further research will determine whether this result can be confirmed or not.

References
Falk, Dan, "Did the Earth Move for You?" *Independent*, 19 June 1998: 10; also available on the Electric Library at <http://www.elibrary.com> accessed 30 April 1999.
Jeanloz, Raymond, and Barbara Romanowicz, "Geophysical Dynamics at the Center of the Earth," *Physics Today*, August 1997: 22–27.
Kanamori, Hiroo, "Shaking without Quaking," *Science*, 27 March 1998: 2063–64.
Ladbury, Ray, "Research Suggests a Recipe for a Lighter Core for Earth—Just Add Water," *Physics Today*, March 1998: 17–19.
Marcus, Steven L., et al., "Detection and Modeling of Nontidal Oceanic Effects on Earth's Rotation Rate," *Science*, 11 September 1998: 1656–58.
Monastersky, Richard, "Ringing Earth's Bell," *Science News*, 4 July 1998: 12–13.
Okuchi, Takuo, "Hydrogen Partitioning into Molten Iron at High Pressure: Implications for Earth's Core," *Science*, 5 December 1997: 1781–84.
Suda, Naoki, Kazunari Nawa, and Yoshio Fukao, "Earth's Background Free Oscillations," *Science*, 27 March 1998: 2089–91.
Wilson, Clark R., "Oceanic Effects on Earth's Rotation Rate," *Science*, 11 September 1998: 1623–24.
Wood, B. J., "Hydrogen: An Important Constituent of the Core?" *Science*, 5 December 1997: 1727.

Further Reading
Allen, Philip A., *Earth Surface Processes*, Malden, MA: Blackwell Science, Inc., 1997.
Lowrie, William, *Fundamentals of Geophysics*, Cambridge: Cambridge University Press, 1997.
Sleep, Norman H., and Kazuya Fujita, *Principles of Geophysics*, Malden, MA: Blackwell Science, Inc., 1997.
Vogel, Shawna, *Naked Earth: The New Geophysics*, New York: Plume Books, 1996.

MISCELLANEOUS

Teleportation

"Beam me up, Scotty!" This phrase, perhaps more than any other in science fiction, has captured the exciting potential of teleportation in the world of the future. *Teleportation* is the process by which a particle or group of particles is transferred, essentially instantaneously, from one point to another. Depending on the context, the term has been used to describe such a transfer by physical or extrasensory means.

For nearly a decade, physicists have realized that this fantastical process may, in fact, be physically possible. That possibility arises out of the laws of quantum mechanics that often predict, allow, or even require behavior that seems completely irrational to human reason.

For example, quantum laws suggest that two particles derived from the same source can undergo a process known as *entanglement*. Entanglement means that the properties of the two particles are inextricably linked, no matter how far they are separated from each other. In a sense, the term suggests the phenomenon in which it is claimed that identical twins can sense what is happening to each other no matter how far or how long they have been separated.

As an illustration, imagine that two photons are created at a single point in such a way that they are entangled. In that case, both photons will have identical properties, such as identical spin and direction of travel. Then suppose that the two photons are separated from each other by some means. For example, they could be directed at a half-silvered mirror that would allow one photon to pass through the mirror and the other to be reflected. The two photons of the entangled pair would still have properties tied to each other even though they have become separated in space.

In such a case, an observer does not know precisely what the properties of the photons that make up this pair are. For example, an observer has no intrinsic knowledge of the direction of spin the particles have. Of course, the researcher can carry out a measurement of that property to determine its quantitative value: for example, to find out the direction in which one of the two photons is spinning.

But the very act of measuring the properties of one particle changes those properties. Determining the spin of a particle, for example, might cause the direction of spin to reverse. In such a case, the other particle in an entangled pair will respond to that change. If the first particle reverses its spin from clockwise to counterclockwise, the second particle simultaneously reverses its spin from clockwise to counterclockwise, no matter how far away the particle is.

Experiments by which the phenomenon of entanglement could be created and tested were suggested in 1992 by Charles H. Bennett of the IBM Thomas J. Watson Research Center in Yorktown Heights, New York, and his colleagues. The problem was to solve a number of critical technical difficulties encountered in designing an experiment to test this theory. In late 1997, two research teams announced almost simultaneously that they had successfully solved these problems and observed teleportation. The two teams were from the University of Innsbruck (in Austria) and the University of Rome.

The Innsbruck experiment illustrates one method by which entangled photons can be used to send a message over a distance instantaneously and without physical contact. In this experiment, a beam of ultraviolet light was passed through a crystal twice. During the first pass, a pair of entangled photons was created. One member of that pair was then sent on to a sending station and the second to a receiving station.

During the second pass of the ultraviolet light through the crystal, another pair of photons was produced. One member of this pair was also sent to the sending station and the second member to a detector that confirmed that the second photon pair had, in fact, been produced. The photon sent to the sending station was passed through a polarizer, which gave the photon a known angle of polarization. This photon then became the "message" the sending station was to send to the receiving station.

At the sending station, the entangled photon and the message photon were combined with each other in such a way that they, too, became entangled. The ability to find a method for entangling these two distinct photons represented a crucial technical breakthrough in the experiment. The leader of the Innsbruck team commented that the team had "learned how to entangle independently created photons," an accomplishment, he said, that would open up "a whole new class of experiments not previously possible" (Peterson, 41).

When the two photons were combined at the sending station, the properties of the entangled photon were detected. As required by quantum theory, this detection process caused the second member of the original entangled pair, detected at the receiving station, to change in response to what had happened to its partner. Detectors at the receiving station were able to "read" this change and, by implication, determine the corresponding property of the member of the entangled pair at the sending station.

The mechanism described here holds little promise for the teleportation of mice, manuscripts, or humans. However, it has other possible applications. For example, it may be an efficient method for transmitting coded messages. A sender is able to produce a message (such as the polarized photon in the Innsbruck experiment) that is to be sent to some second party. That message can then be encoded by combining it with entangled particles sent from some third party. Neither the sender nor the receiver needs to know what the nature of the entangled particle is. All they need to know is that whatever happens to the message at the sending station when it is combined with the entangled particle, the reverse will occur at the receiving station. The receiving station only has to subtract the effect of the entangling particle from the message received to know what the original message was.

Research on teleportation by means of entangled particles has moved forward very rapidly. Shortly after the Innsbruck and Rome experiments were reported, similar successes in the entanglement of electrons and kaons were reported. The kaon research was particularly significant because kaons are relatively massive particles compared to photons (which have no mass) and electrons (which have very little mass).

In the kaon experiment, a particle accelerator at the Centre Européen pour la Recherche Nucléaire (CERN) was used to slam a beam of antiprotons into a target of hydrogen gas. One of the results of this very complex reaction was the formation of pairs of kaons and antikaons. A kaon is a particle that consists of one up quark combined with one strange quark.

When kaons and antikaons were formed in the target gas, they escaped in opposite directions. CERN researchers were able to show that measuring the properties of one beam of escaping particles affected the properties of the other beam. For example, when they detected the presence of a kaon in one part of the detection device, thus "forcing" the particle to identify itself as a kaon, the comparable beam in a second part of the detector "revealed" itself to consist of antikaons. Thus, the actions taking place in one region of space influenced what was taking place at the same time in a different region of space a few centimeters distant from the first location.

References

"Austrian Scientists 'Beam' Matter across Their Lab," *San Francisco Chronicle*, 12 December 1997: A13.

Collins, Graham P., "Quantum Teleportation Channels Opened in Rome and Innsbruck," *Physics Today*, February 1998: 18–22.

Peterson, Ivars, "Instant Transport," *Science News*, 17 January 1998: 41.

"Quantum Teleportation" <http://www.research.ibm.com/quantuminfo/teleportation/> accessed 30 April 1999.

"A Quantum Wedding of Quarks" <http://sciencenow.sciencemag.org/cgi/content/full/1998/60413> accessed 4 June 1999.

Order from Chaos

Entropy is a fundamental phenomenon in nature. Entropy is the process by which any system becomes more disordered. As an example, picture a sugar cube dropped into a glass of water. Over time, the highly ordered structure of the sugar cube breaks down as individual sugar molecules are dissolved and carried away into solution. Ultimately, the sugar cube disappears completely to be replaced by a homogeneous, completely disordered solution of sugar water.

Scientists believe that disorder is the final fate of the universe. Over untold periods of time, all physical systems become more and more disordered until entropy finally reigns supreme.

Physicists have been shocked in recent years, therefore, to discover that some systems exhibit a decrease in entropy, with ordered structures building up out of disordered states. These discoveries are somewhat akin to watching a crystal of sugar begin to assemble itself spontaneously from a solution of sugar water.

One example of this phenomenon was first reported in 1992 by Peter Kaplan, then a graduate student at the University of Pennsylvania. Kaplan was studying the behavior of microscopic spheres of two different sizes added to a saline (salt) solution. Entropy predicts that the two types of spheres will distribute themselves randomly and homogeneously in the saline solution. In this condition, entropy in the system would be maximized.

However, Kaplan's mixture did not display this behavior. Instead, larger spheres began to collect in one corner of the container, producing ordered, crystalline-like structures. Out of random movement had come order.

Kaplan's advisor, Arjun Yodh, and his colleagues have continued to study this behavior since 1992. They have discovered that the ordering of particles from a random mixture tends to follow certain specific rules. For example, they have learned that ordering in a cylindrical tube is most likely to take place along the edges of the tube that are most bent.

Kaplan and Yodh developed an explanation for this seeming violation of the principle of entropy. They pointed out that entropy did decrease with regard to the formation of large-sphere "crystals." However, by taking these larger spheres out of circulation, as it were, more room was available for smaller spheres to spread out. That is, entropy among the smaller spheres actually increased at the expense of the larger spheres. Taken as a complete system, entropy continued to increase, although it did decrease in one component of the system—the part occupied by the larger spheres.

Formation of Electron "Crystals"

A second example of the "defeat" of entropy was reported in 1994 by researchers at the University of California at San Diego (UCSD) under the direction of Fred Driscoll. In this case, the system being analyzed consisted of groups of electrons trapped in a vacuum and surrounded by magnetic fields. Under this condition, a mass of electrons is essentially

Electrons in a vacuum form an ordered pattern, an apparent contradiction to the expected increase in entropy in the system.
Courtesy of C. Fred Driscoll, UCSD.

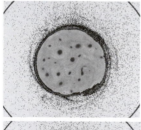

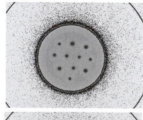

free to travel randomly throughout the container, constrained only by the exterior magnetic force. One would expect to see the development of a completely random, homogeneous blur of electrons throughout the container.

To test this prediction, Driscoll's team recorded the location of electrons in the container on a phosphorescent screen. The smooth distribution of light expected because of entropy did not occur, however. Instead, the team found electrons collecting in discrete whirling vortices that slowly grew larger over time. At some point, these vortices took on a fixed structure that looked something like a crystal made of electrons. As in the Kaplan-Yodh experiment, ordered structures had spontaneously grown out of a system in which one would have expected to see increased entropy and randomness.

The explanation developed for this phenomenon by Daniel Dubin and Dezhe Jin, also at UCSD, is similar to that devised for the Kaplan-Yodh observation. Entropy obviously does decrease within the electron "crystals" observed by Driscoll's team. But the drawing together of some electrons into highly ordered structures leaves electrons outside the vortices free to move with even greater freedom than before. Again, the apparent "defeat" of entropy in one small part of the system is more than compensated for by the increase in entropy in other parts of the system. From the broadest view of the system, entropy continues to reign supreme, as it must finally do anywhere in the final scheme of nature.

References

Dinsmore, A. D., et al., "Hard Spheres in Vesicles: Curvature-Induced Forces and Particle-Induced Curvature," *Physical Review Letters*, 12 January 1998: 409–12.

Glanz, James, "From a Turbulent Maelstrom, Order," *Science*, 24 April 1998: 519.

Jin, D. Z., and D. H. E. Dubin, "Two-Dimensional Vortex Crystals," *New York Academy of Sciences*, January 1998: 18–21.

Kestenbaum, David, "Gentle Force of Entropy Bridges Disciplines," *Science*, 29 March 1998: 1849.

Further Reading

Gleick, James, *Chaos: Making a New Science*, New York: Penguin USA, 1988.

CHAPTER TWO
Applied Research in Physics

INTRODUCTION

The studies reported in this chapter illustrate the intricate interplay between basic and applied research. All of the studies draw, of course, on basic research that has provided fundamental information about the nature of matter and energy. These studies of applied physics illustrate the ways in which that basic information can be put to use for practical purposes.

But the studies also show how applied research can result in new discoveries or raise new questions for basic research. Thus, solving one problem does not represent the end of a research line, but rather is a way station in learning about and wondering about related aspects of nature.

This chapter provides only a small selection of the many exciting discoveries announced in applied physical research during the late 1990s. These reports do offer, however, a hint of the excitement and advances that characterize many other research developments during the same period.

ACOUSTICS

Sound is one of the most common phenomena in everyday life, so it may seem surprising that physicists are still making some intriguing breakthroughs in our understanding of this very basic form of energy. Yet, such has been the case in the past decade. Some of the most interesting of the new discoveries about sound are described below.

Resonant Macrosonic Synthesis (RMS)

Beginning physics students learn that energy can occur in many forms: kinetic, nuclear, electrical, magnetic, and so on. Sound is always listed as a form of energy, but it may seem a stretch to think of sound in this context. If energy is the ability to do work, how does sound fit into this classification?

One problem that limits the use of sound as a source of energy is the phenomenon of shock waves. Many individuals are familiar with the booming sound of a shock wave created when a jet airplane breaks through the sound barrier. The problem in using sound waves for work is that they are typically able to absorb only a certain given amount of energy. Energy (for example, kinetic energy) can be pumped into a sound wave up to some critical point. However, if an attempt is made to add more energy beyond that critical point, that energy is not added to the wave, but is "packaged" as a shock wave and released into the surrounding environment. The total energy present in the sound wave is thereby reduced to some value below the critical point.

In practical terms, the assumption has always been that a sound wave can be made only so loud. Increasing the volume (energy) of the sound beyond some critical point is just not possible. But without a louder (more energetic) sound, the wave cannot be used to perform useful work.

In 1997, researchers at the MacroSonix Corporation in Richmond, Virginia, announced a breakthrough in the production of very loud (energetic) sounds. They called their system resonant macrosonic synthesis (RMS). The key to this development was the discovery that the shape of a container is critical in the formation of a shock wave. When efforts are made to produce high-energy sound waves in a cylinder, as described above, a shock wave inevitably forms. Because of this, MacroSonix researchers investigated the possibility of generating sound waves in containers of unusual shapes. The shape they found to be most effective is shown in the illustration below. A sound wave can be generated in this container by shaking it back and forth along its axis. The kinetic energy added through shaking continuously increases the amount of energy carried by the sound wave generated inside the container, but the shape of the container apparently prevents a shock wave from forming. The total energy carried by a wave in such a container continues to build until it becomes significantly greater than that produced by any sound wave in the past.

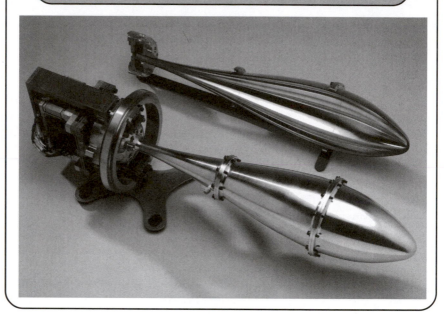

An RMS system: A solid resonator with drive motor (foreground), shown also in a cutaway view (background). *Courtesy of MacroSonix Corp.*

MacroSonix researchers reported the maximum sound produced in this container was about 200 decibels. A decibel is the unit used to measure sound intensity. A jet engine produces about 150 decibels, and a sound of 165 decibels has sufficient energy to set fire to a person's hair. Since the decibel scale is a logarithmic scale, a decibel rating of 200 is about 3,000 times greater than that of the fire-setting level.

Acoustics experts can imagine a number of practical applications for the MacroSonix invention. Any device that makes use of a compressor, for example, could employ an acoustic compressor rather than one of the gas compressors currently in use. The acoustic compressor has almost no moving parts and, therefore, should be simpler in design and construction, less expensive, and easier to maintain. The most obvious first application for the acoustical compressor might be in home refrigerators. In fact, MacroSonix has announced that one major appliance manufacturer has already made a licensing arrangement to build a refrigerator based on the acoustic compressor.

But the acoustic compressor has many other potential uses also. For example, the sound waves it generates are powerful enough to keep heavy objects levitated (floating) in air. The use of sound waves to levitate Ping-Pong balls has often been used as a demonstration by physics instructors. According to MacroSonix president and CEO Tim Lucas, however, the RMS technology "should allow us to levitate bowling balls with sound waves" (Ashley, 1). MacroSonix's Web page is at <http://www.macrosonix.com>.

References
Ashley, Steven, "Sound Waves at Work" <http://www.macrosonix.com/030098me.htm> accessed 30 April 1999.
Ladbury, Ray, "Ultrahigh-Energy Sound Waves Promise New Technologies," *Physics Today*, February 1998: 23–24.
Mackenzie, Dana, "Cool Sounds at 200 Decibels," *Science*, 19 December 1997: 2060.
Peterson, I., "Breaking through the Acoustic Shock Barrier," *Science News*, 6 December 1997: 358.
"Scientific Discovery Leads to Breakthrough Technology" <http://www.macrosonix.com/ms120197.htm> accessed 19 October 1998.

Acoustic Tomography

Interest in the Earth's oceans has increased rather significantly in the past two decades. One reason for this increased attention has been concerns about global climate change. As evidence begins to accumulate that the Earth's climate is becoming warmer, scientists want to learn more about the role of oceans in any such long-term change. The oceans and the atmosphere are, after all, the most important components of the complex system that determines the Earth's climate.

Studying the oceans requires techniques somewhat different from those used in studying the atmosphere, however. For example, one can see vast distances into the atmosphere with cameras and other devices that measure light. But the oceans are largely opaque to visible light. To "see" what is going on in the oceans, other techniques need to be employed.

In 1979, Walter Munk at the Scripps Institution of Oceanography and Carl Wunsch at the Massachusetts Institute of Technology suggested using sound waves to study the oceans. They proposed using acoustical tomography to determine the temperature and heat content of ocean water over short periods of time. Their plan was to obtain real-time (immediate) measurements of such characteristics, rather than waiting for long-term studies that measure such properties over extended periods of time.

Tomography is a method for determining the characteristics of a three-dimensional object by passing waves through that object. The medical examination techniques known as CAT scans and PET scans are everyday examples of this technology. In each case, waves (X rays in the case of CAT scans and positrons in the case of PET scans) are passed through the body. Changes that take place in the waves as they move through the body provide detailed information about the structure of the region being studied. The interpretation of the actual picture produced in either case is not a simple matter, but techniques have now been developed that allow the effective everyday use of both techniques.

Acoustic tomography operates on the same principle as CAT and PET scans, except that sound waves are used rather than X rays or positrons. Interest in the use of acoustic tomography for ocean studies developed as scientists became more concerned about finding changes in temperature patterns across the broad stretch of the oceans. The program that was developed to carry out this research was given the name Acoustic Thermometry of Ocean Climate (ATOC).

ATOC was scheduled to begin operation in 1994. The plan was to install sound sources (loudspeakers) on the Pioneer Seamount, an underwater structure about 88 kilometers west of Pillar Point, California. The source was designed to produce 260 watts of acoustic power (195 decibels, a water standard). Microphones for the reception of these sound waves were installed near Hawaii, Christmas Island, and New Zealand, as well as provided by the U.S. Navy at a number of locations in the North Pacific.

ATOC source. *Courtesy of James Mercer, APL-UW.*

The ATOC program did not begin on time, however. Some environmentalists were concerned that marine animals would be hurt or killed by the sounds. They asked for a delay in the start of ATOC to study these effects.

ATOC researchers acceded to this request and devoted all of the first year of their experiment to testing the effects of sound waves on marine animals. At the end of that period, environmental biologists and everyone involved with the ATOC project were convinced that no harm had come to any animals. The determination was made to go ahead with the originally planned research.

One year later, the ATOC research team reported the results of its first year of research. They announced that the technique was even more effective than had been expected. Sound waves sent out from the Pioneer Seamount were detected as far as 5,000 kilometers away. The sounds could be detected with a precision of about 20–30 milliseconds after the sound waves had been traveling for nearly an hour through the ocean. With this degree of precision, it should be possible to detect temperature changes in the ocean with a precision of about 0.006 degrees celsius.

Two specific conclusions from this research were of special interest. First, sound wave patterns made it possible to measure tidal movements with great accuracy. When those measurements were compared with data on tidal patterns collected by other means (such as direct observation), they were found to be in close agreement. Our present understanding of the behavior of tides can, therefore, be accepted as reasonably accurate.

The second research finding involved changes in total ocean volume as a result of heating. As the oceans absorb solar radiation they become warmer, of course, and tend to expand in volume. Measurements taken by the TOPEX/POSEIDON satellite had earlier provided very precise measurements of increases in sea level and, hence, ocean volume. Data collected from the ATOC study found, however, that temperature changes measured by acoustic tomography could account for only about half of this increase in volume. Oceanographers now face the task of finding an explanation for the remaining increase in water volume measured by TOPEX/POSEIDON.

The ATOC experiment was designed to continue through the end of 1999. At that point, the original grants for the research will have expired. As of late 1998, researchers were still "cautiously optimistic" about finding new funding for their project.

References
"Acoustic Monitoring of the Ocean" <http://www1.etl.noaa.gov/acoust/acoust.htm> accessed 30 April 1999.

ATOC Consortium, "Ocean Climate Change: Comparison of Acoustic Tomography, Satellite Altimetry, and Modeling," *Science,* 28 August 1998: 1327–30.

Bunch, Bryan, *Handbook of Current Science & Technology,* Detroit: Gale, 1996, 540.

Howard, J., "Listening to the Ocean's Temperature," *Explorations* 5 (2), Fall 1998; also <http://www-atoc.ucsd.edu/Explorations_f98/Atoc.html>.

Monastersky, R., "A Sound Way to Take the Sea's Temperature?" *Science News,* 29 August 1998: 133.

Munk, W., P. Worcester, and C. Wunsch, *Ocean Acoustic Tomography,* New York: Cambridge University Press, 1995.

"Ocean Acoustic Tomography" <http://www.oal.whoi.edu/tomo2.html> accessed 30 April 1999.

Taroudakis, Michael, "Inversion Algorithms for Ocean Acoustic Tomography" <http://www.ercim.org/publication/Ercim_News/enw22/inversion-algorithms.html> accessed 30 April 1999.

Worcester, Peter, "ATOC Thermometry: Results from Over a Year of Acoustic Transmissions" <http://atoc.ucsd.edu/ATOCupdatepg.html> 30 April 1999.

Acoustical Tenderizing of Meats

Advances in chemistry and biology often can be applied to the food sciences. For example, a new chemical compound may be discovered that mimics a natural flavor or odor, or that retards the spoiling of foods. It is less common to hear about an advance in physics that can be applied to improving our food supply. In 1998, however, just such an advance was announced.

That development involves the use of high-intensity acoustical waves to tenderize meats. The method was invented by John B. Long, a mechanical engineer retired from the Lawrence Livermore National Laboratory (LLNL). Long had worked on explosives research at LLNL and had become interested in the question of how research of that type could be used for peacetime applications.

It occurred to him that one application might be in the tenderizing of meats. Long was aware that a significant portion of most food animals, such as cattle and sheep, are classified as "select" because they are less satisfactory than cuts classified as "prime" or "choice." Select cuts are generally tougher and, therefore, less desirable. The major difference between select cuts and prime and choice cuts is fat content. Select and prime cuts tend to have more fat intermixed with muscle tissue. The presence of fat makes meats juicier and easier to cut and chew. The disadvantage of fat-marbled meats, of course, is that they tend to be less healthful because of their fat content.

Long hypothesized that shock waves sent through a piece of meat would break down muscle fiber, resulting in a product that is more tender and, therefore, more attractive to consumers—but with a lower fat

In this apparatus, meat is tenderized using sound waves from a high-energy explosive charge. *Courtesy of USDA-ARS.*

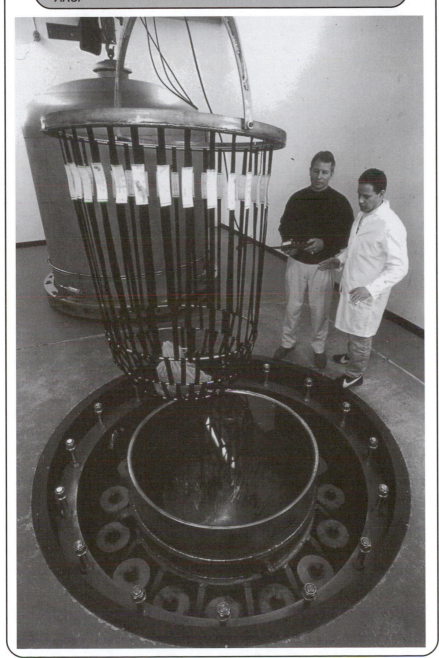

content than prime or choice grades of meat. His first experimental apparatus designed to test this hypothesis consisted of a plastic drum covered with a steel plate and filled with water. Cuts of meat were wrapped in watertight material and submerged in the water. Then, a small charge of explosive was set off inside the drum. Initial tests of this kind were convincing enough that the U.S. Department of Agriculture became interested in more extensive tests of the procedure. Long was also able to find partners willing to invest money in his scheme as a commercial venture. The company that resulted from that investment is now known as Hydrodyne, located in Buena Vista, Virginia.

A prototype tenderizing unit being tested at Buena Vista consists of a steel tank weighing 7,000 pounds and covered with a steel dome eight feet in diameter and weighing 5,000 pounds. Packages of meat weighing up to 600 pounds can be treated in a single step by means of a small explosive set off two feet above the meat. The shock wave produced in this explosion is reflected over the walls and dome of the tank and produces a pressure as high as 40,000 pounds per square inch.

Meats treated by this process have been judged by tasting panels as being at least as tender as prime and choice cuts, without any negative effects on their color or shelf life. Long's company believes that the acoustic tenderizing process will significantly increase the economic use of meats of all kinds from all parts of an animal.

References

Lee, Jill, "Big Shock Makes Tender Beef" <http://www.ars.usda.gov/is/pr/1998/980629.htm> accessed 30 April 1999.

Lee, Jill, "Hydrodyne Exploding Meat Tenderness," *Agricultural Research,* June 1998; also <http://www.ars.usda.gov/is/AR/archive/jun98/hydro0698.htm> accessed 31 May 1999.

Raloff, Janet, "Ka-Boom!" *Science News,* 6 June 1998: 366–67.

Further Reading

Kinsler, Lawrence E., *Fundamentals of Acoustics,* 3rd edition, New York: John Wiley and Sons, 1982.

Medwin, Herman, and Clarence Samuel Clay, *Fundamentals of Acoustical Oceanography,* New York: Academic Press, 1997.

Urick, Robert J., *Principles of Underwater Sound,* Los Altos, CA: Peninsula Publishing, 1996.

NANOTUBES

Care for a glimpse at physics in the twenty-first century? If so, a brief review of nanotubes is what you want.

Nanotubes are essentially cylindrical structures with diameters of no more than a few nanometers (billionths of a meter), but with lengths of up to 100 micrometers (millionths of a meter). They are the ultimate in long, thin, hollow bits of matter.

Nanotubes were first discovered in 1991 by the Japanese electron microscopist Sumio Iijima. They can be considered to be derivatives of fullerenes, an allotrope of carbon discovered in 1985 by Richard E. Smalley, Robert Curl, and Harold Kroto. The first fullerene discovered was a molecule consisting of 60 carbon atoms joined to each other in the shape of a geodesic dome. The molecule was named buckminsterfullerene, in honor of Buckminster Fuller, the man who made geodesic domes popular as an architectural form in the 1950s. These molecules are now known as *buckyballs*. Other large structures of carbon atoms have since been discovered and are known more generally as *fullerenes*. The basic unit of a fullerene molecule is a ring consisting of five, six, or seven carbon atoms. These pentagons, hexagons, and heptagons are the building blocks from which fullerene molecules consisting of 60, 70, or some other large number of carbon atoms are constructed.

Buckyballs and Nanotubes

To better understand the relationship of fullerenes to nanotubes, imagine a single 60-carbon-atom buckyball sliced down the middle. Then imagine the two 30-carbon fragments formed joined by a circular strip of carbon atoms also joined in hexagonal or pentagonal patterns. The molecule formed in this way is a short cylinder, one atom thick throughout, with hemispherical caps on either end. By adding more and more circular strips at the center of the molecule, the cylinder can be made any length desired. A nanotube is simply a molecule of this type with many thousands of circular strips inserted between two hemispherical caps.

Fortunately for scientists, the steps involved in carrying out this "thought experiment" to make a nanotube is far more complex than what occurs in the real world. In fact, scientists now know that nanotubes form spontaneously and naturally whenever carbon is burned in a limited supply of oxygen. The soot that forms under such conditions often consists of fullerene and nanotube molecules.

A better way of understanding nanotubes may be by comparison with the familiar form of carbon known as graphite. Graphite is the slippery black material used as a solid lubricant and as the "lead" in an ordinary pencil. Its molecular structure consists of a flat plane of carbon atoms joined in a maze of hexagons. Each carbon atom in the maze shares three of its four valence electrons with three other carbon atoms around it. The fourth valence electron remains unbonded—"hanging loose," as it were—above or below the plane of the graphite sheet.

The molecular structure of graphite is, therefore, different from that of diamond and nearly all organic compounds, in which all four valence electrons are used to form four bonds with other atoms around it. Pure diamond, for example, consists of carbon atoms, each of which is bonded to four other carbon atoms adjacent to it.

Because of its molecular structure, graphite exhibits a particularly interesting property when it is heated. The edges of a graphite sheet contain loose or "dangling" bonds, consisting of the single unpaired electron from each carbon atom. These single electrons have a strong tendency to seek out and pair with other such electrons. In a practical sense, what this means is that a flat graphite sheet tends to curl when it is heated. The curling occurs because "dangling" bonds on all edges seek similar bonds on other edges with which to join.

Simply stated, nanotubes can be produced by the simple process of heating graphite. As heating occurs, sheets of graphite curl around a central axis and join along their edges, forming long, thin, cylindrical structures of nanotubes.

More Complex Nanotubes

Scientists have now learned how to engineer the precise form of a nanotube by adjusting temperature and other conditions under which a sample of carbon is heated. They have also learned how to introduce other elements into a nanotube such that atoms of boron, nitrogen, and other elements can take the place of carbon.

Nanotubes are now generally classified as single-wall nanotubes (SWNT) or multiple-wall nanotubes (MWNT). SWNT have structures like those described above; that is, they are no more than one atom thick. MWNT, by contrast, consist of many one-atom-thick tubes nested within each other. One form of MWNT consists of a single SWNT composed entirely of carbon atoms deposited in the middle of two or more other SWNTs in which one or more non-carbon element has been added.

The various forms of nanotubes have, as one might predict, differences in physical characteristics, such as in strength, flexibility, and electrical and thermal conductivity.

Properties of Nanotubes

Nanotubes have a number of remarkable physical properties that suggest they may be useful for revolutionary applications in the future. For example, they are the strongest material ever discovered. They are 100 times stronger than the best steel and are stronger than comparable rods made of silicon carbide (SiC), one of the toughest materials known. Nanotubes also maintain their strength after bending, twisting, and buckling. Richard Smalley has described nanotubes as "the stiffest stuff you can make out of anything" (Service, 942).

In some regards, the most interesting property of nanotubes at this point in time is their electrical conductivity. Carbon itself is a good conductor of electricity, and so are nanotubes in their most familiar form. That fact in and of itself is not particularly surprising. What *is* surprising is that the conductivity of nanotubes can be altered rather easily simply by changing their diameters and/or their axial alignment.

In 1997, for example, Alex Zettl at the University of California, Berkeley, discovered that a simple defect in the wall of a nanotube causes a dramatic change in electrical properties. In some cases, the normal all-hexagon structure of a nanotube is replaced by a pair of rings, one a pentagon and the other a heptagon. When a "defect" of this kind is present, the shape of the nanotube changes. Instead of having all hexagons arranged in a perfectly straight line along the length of the tube, a slight twist is created. The hexagons are then arranged in a spiral pattern around the vertical axis of the tube.

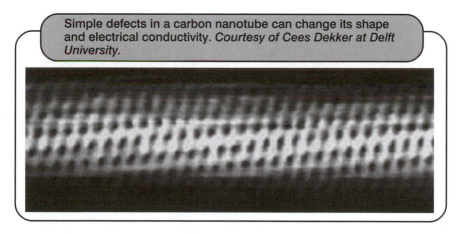

Simple defects in a carbon nanotube can change its shape and electrical conductivity. *Courtesy of Cees Dekker at Delft University.*

Although that modification might appear to be a modest difference, it has a profound effect on the electrical properties of the tube—that is, in the parts of the tube where the hexagons are aligned in a straight line, the nanotube conducts electrical current very well, as does graphite. In regions where the hexagons are arranged in a spiral pattern, however, the tube becomes a semiconductor. It conducts an electrical current under some conditions, but not others.

Scientists have also discovered that nanotubes in the "normal" (straight-hexagon) arrangement appear to conduct an electric current almost without resistance. While they are not superconductors in the traditional sense, they do permit the flow of electricity virtually without resistance at room temperatures. An electrical device made out of nanotubes could, therefore, be expected to operate with very high efficiency.

Other fascinating properties of nanotubes are constantly being discovered. For example, in 1998, researchers at the Swiss Federal Institute of Technology reported that they had observed luminescence from nanotubes through which an electrical current was being passed. When electricity was passed through the nanotubes, they gave off a very faint—but easily visible—green glow. The Swiss researchers explained the luminescence in this way: Electrons flow easily through the long chain of carbon atoms in a nanotube. All electrons in this part of the molecule are at the same energy level, so no energy is gained or lost and, hence, no radiation is absorbed or emitted. At the end of the tube, however, discrete energy levels do exist in the hemispherical buckyball cap. An electron is able to jump from a lower energy level to a higher energy level or from a higher level to a lower level. In the latter case, a small amount of light will be given off. It is this tiny amount of light, the researchers believe, that can be observed during the luminescence of the nanotube.

Applications of Nanotubes

Nanotube research illustrates the amazing speed with which fundamental discoveries in physics can find important practical applications. For example, a nanotube with a pentagon-heptagon "defect" could be used as a semiconductor that allows current to pass in only one direction. In that context, it would be the smallest rectifier or diode ever constructed. The device would be particularly useful since a single nanotube molecule could be used both as an electrical conductor in regions where the hexagon pattern is undisturbed and as a semiconductor in the "defect" region.

Some very simple electrical devices have already been constructed with nanotubes. In May 1998, for example, Cees Dekker and his colleagues at the Delft University of Technology in the Netherlands described the use of a single nanotube as a transistor. In this experiment, Dekker laid down two small gold electrodes on a layer of silicon oxide that covered a base of silicon. A single semiconducting nanotube was then laid across the two gold electrodes. When a current was applied to the silicon base, electrical current flowed through the nanotube, from one electrode to the other.

Other applications will be based on the strength of nanotubes. For example, material scientists predict that nanotubes will be used to make composites that are stronger and lighter than any existing material. One such application, albeit a very specific and exotic application, is in the production of probes for scanning tunneling microscopes. Such microscopes are used to study atom-sized detail on surfaces. The probes they

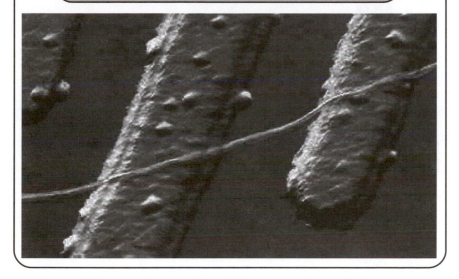

A carbon nanotube lying across two metal electrodes can act as a transistor. *Courtesy of Cees Dekker at Delft University.*

contain must be very small and very strong. The experimental use of nanotube composites in the production of such probes has already met with some success, and this limited application has already been realized.

Another suggested use of nanotubes is as circuitry in flat-panel televisions. Such devices have been constructed and tested, but they require a very different arrangement of electronic components in order to reduce their depth to a matter of no more than a few inches. The Ise Electronic Corporation has already announced plans to use nanotubes in the circuitry of flat-panel televisions that they hope to have commercially available shortly after 2000.

Future Directions

Variations on traditional nanotube research are occurring at a bewildering rate—even though "traditional nanotube research" is itself less than a decade old! For example, researchers are exploring the effects of filling the core of a nanotube with other materials, such as silver or aluminum, in order to produce a carbon-coated wire only one atom thick. Such a device would be likely to have properties and applications that neither a larger wire of the same material nor an empty nanotube has.

Carbon nanotubes are also being used as templates in the manufacture of nanowires of other materials. For example, one experiment described in 1997 involved the heating of carbon nanotubes located on a cathode in connection with a hafnium boride (HfB_2) anode in an atmosphere of pure nitrogen. The product of this reaction was a MWNT consisting of alternate layers of pure carbon and boron nitride with properties distinct from those of a pure carbon nanotube.

References

Collins, Philip G., et al., "Nanotube Nanodevice," *Science*, 3 October 1997: 100–03.

Fasol, Gerhard, "Nanowires: Small Is Beautiful," *Science*, 24 April 1998: 545–46.

Perkins, S., "Nanotubes: Metallic by a Twist of Fate," *Science News*, 10 January 1998: 22.

Peterson, I., "Basing Transistors on Lone Carbon Nanotubes," *Science News*, 9 May 1998: 294.

Saito, Susumu, "Carbon Nanotubes for Next-Generation Electronics Devices," *Science*, 3 October 1997: 77–78.

Service, Robert F., "Superstrong Nanotubes Show They Are Smart Too," *Science*, 14 August 1998: 940–42.

Suenaga, K., et al., "Synthesis of Nanoparticles and Nanotubes with Well-Separated Layers of Boron Nitride and Carbon," *Science*, 24 October 1997: 653–55.

Wu, C., "Nanotubes Get Another Glowing Review," *Science News*, 22 August 1998: 116.

Further Reading

Drexler, K. Eric, *Engines of Creation*, New York: Doubleday, 1986.

Harris, Peter J. F., *Carbon Nanotubes and Related Structures: New Materials for the 21st Century*, Cambridge: Cambridge University Press, 1999.

Regis, Ed, *Nano: The Emerging Science of Nanotechnology*, Boston: Little Brown and Company, 1995.

ten Wolde, Arthur, *Nanotechnology: Towards a Molecular Construction Kit*, The Hague: SST (Netherlands Study Center for Technology Trends), 1998; also <http://www.stt.nl/_privateE/hpnan.htm>.

NANOTECHNOLOGY

Have you ever had the feeling that engineers seem to announce the development of a newer and faster computer chip every year or two? If so, you're not far off. In fact, Gordon Moore of Fairchild Semiconductors predicted in 1965 that the number of transistors engineers could pack on a computer chip would double about every 18 months. Moore's prediction, now known as Moore's Law, has turned out to be remarkably correct. While the rate in development of processing chips has fallen somewhat behind the rate predicted by Moore, the rate of development for memory chips has exceeded that rate by a nearly comparable amount.

Moore's Law is more than a prediction. It also carries an inherent warning for engineers: Progress in reducing chip size cannot continue indefinitely. At some point, improvements in chip circuitry will be limited by fundamental physical constraints. One can obviously not make smaller and smaller circuitry when the dimensions of "smaller" become those of atoms and molecules.

What happens, then, as we approach these physical limitations in the early twenty-first century, when Moore's Law predicts that chip circuitry can no longer be made any smaller?

At that point, the challenges facing engineers are twofold. The first of these challenges is purely technological. It involves the development of methods for making components smaller and smaller. Enormous technological advances were necessary in order to find ways of etching patterns in silicon and other semiconducting materials and laying down circuits with wires no more than a fraction of a micrometer in diameter. But the tools used today for such processes will not be useable for components that are one half or one quarter of those dimensions, or less. Completely new technologies will have to be developed.

A second challenge is more fundamental. The new technologies needed for the manufacture of nanometer-size circuits will have to operate at the multiatom or multimolecule (or even the single-atom and single-molecule) level. At this level, the physical principles with which engineers have to deal change dramatically. The properties and behavior of materials in this region are controlled by the principles of quantum mechanics rather than by those of classical physics.

For example, it may no longer be possible to specify that a particular particle occupies a specific position in space. Instead, it may occupy—or may even be required to occupy—two different positions at the same time. It is difficult to imagine what a computer chip produced in 2025 will look like, given these principles that are so strikingly at variance with those we are accustomed to dealing with in everyday life.

Scientists and engineers have been aware of these two challenges for some time. Dozens of research teams around the world have been investigating many of the theoretical and practical problems that have arisen (and will arise) in the new era of nanotechnology. Although space limitations preclude a comprehensive and detailed description of the advances being made in this field, some examples of the kinds of progress being made can be reviewed.

The Smallest Metal-Oxide Semiconductor

Warnings about the limitations of existing transistor designs have already begun to appear in the research literature. An example is the work of Steven Hillenius and his colleagues at Bell Laboratories. In 1998, the Bell team reported on the construction of the smallest transistor ever made. The transistor proved itself to be the most efficient transistor ever constructed, but its operation posed some serious issues for the Bell team and others involved in the development of transistors.

The tiny transistor developed at Bell consisted of the standard components: a conducting base of silicon doped with other atoms and covered with an insulating layer of silicon dioxide, to which is attached a narrow gate that controls the flow of electric current through the transistor. In the Bell design, the gate was 60 nanometers wide, approximately the size of 180 atoms laid side by side. The insulating layer of silicon dioxide was 1.2 nanometers thick, the size of about three atoms.

The good news about the new Bell transistor was that it delivered a current flow five times faster than today's transistor while drawing 60 to 160 times less power. Thus, the new design represented yet one more step forward in the race to build ever-more-efficient transistors.

The problem was that the Bell transistor appears to have reached the limits posed by current design concepts. For one thing, tunneling of electrons through the insulating layer was observed. Tunneling is a quantum mechanical process by which electrons pass from one point to another through a barrier that would be regarded as sufficient to block them according to the laws of classical physics. As components of a transistor (or any other device) become smaller and smaller, quantum mechanical effects such as tunneling become more and more important. And physicists are not yet certain how such effects will affect the operation of the transistor or the components of which it is made.

As an example of problems that remain to be

An experimental "nanotransistor," only 182 atoms wide. *Courtesy of Lucent Technologies.*

61nm

Tungsten-Silicide

Polysilicon

Drain   Source

solved, the Bell team observed an unexpected phenomenon when the insulating layer was reduced to less than 1.2 nanometers in thickness. Under such conditions, current flow in the device began to decrease as the voltage applied to it was increased. But this phenomenon is just the reverse of what normally happens in a transistor, and on which transistor operation is based. In other words, increasing the voltage applied to the gate normally increases the flow of electrons in a transistor. The Bell team has not yet developed an explanation for this unexpected result, but it appears to pose a serious potential threat to the continued miniaturization of electronic components by traditional methods.

A Single-Electron Transistor

Theoretically, the simplest transistor that one could imagine would consist of a device that allows the flow of electrons one at a time. It will probably be some time before the practical problems involved in the creation of such a device are worked out; however, some initial steps have already been taken in its development. In fact, a team of U.S. and Swedish researchers reported in 1998 on the design of a simple single-electron transistor (SET).

At this point, the U.S.-Swedish device is of interest not primarily because of its transistor applications, but because of its potential for measuring the smallest possible flow of electrical current: the flow of single electrons through a circuit.

Here's how the device works. The conducting and insulating layers of the traditional transistor are combined into a single structure that makes up the base of the SET. The conducting and insulating layers are replaced with a thin layer of insulating material into which is inserted a short piece of semiconducting material. The gate that controls the flow of current through the transistor is attached above the semiconducting portion of the base.

In this arrangement, electrons travel from one end of the base, through the insulator, across the island, and into the opposite end of the insulator under the control of the voltage applied to the gate. When the semiconducting section already contains an electron, no current will flow since electrons in the insulating portion of the base are repelled by it. As voltage is applied to the semiconducting region, however, an electron present within that region will be forced out, allowing another electron from the insulating region to move into the semiconducting region. By controlling the voltage applied to the gate, the flow of electrons through the semiconducting region can be very carefully regulated.

This same arrangement can be used to detect the microscopic flow of current through a wire. Suppose that the wire is laid down next to the base of the SET described above. When a current flows through the wire, it will generate its own electromagnetic field. That field will act on the SET in much the same way as voltage applied to the gate in the tiny transistor. It will allow the flow of electrons through the semiconducting layer of the SET base at a rate that reflects the intensity of the flow in the wire itself. As the current increases, it generates a greater electromagnetic force and a greater flow of electrons through the SET. The advantage of the SET described here, of course, is that it can detect very small currents with high precision.

The U.S.-Swedish team was faced with technical difficulties in finding ways to amplify the very small flow of electrons through the SET. They eventually found a way of converting the frequencies generated by electron movement in the SET to electromagnetic pulses that could be magnified and read. They anticipate that the system they have developed will eventually be able to detect the movement of single electrons through the device.

Wiring Metal Dots

One of the fundamental problems in electronic technology today is finding efficient ways to wire two metallic points to each other. The two points might be an input-output combination, for example, with the wire serving as a connection between the two. Efficient wiring is essential because it is one of the factors that determines the speed with which electrical signals can travel between two points and through a transistor or chip. As engineers look for new ways to improve the specifications of their chips, better methods of wiring become more and more in demand.

In 1997, scientists at Drexel University reported on an interesting new technique for connecting two metallic dots. The traditional method for making such connections is photolithography. In photolithography, a light beam is used to carve a channel into a substrate. The connecting wire can then be laid down into that channel. This process—however easy it is to describe—actually involves a number of steps that make it somewhat inefficient.

The Drexel experiment was one of a long series of efforts to find alternative ways of connecting two dots without using photolithography. One line of that research has involved "growing" a wire between two metallic points by attaching the points to an electric circuit. In the Drexel variation, however, the two metal points were kept physically separate from each other while a wire between them was produced.

The Drexel researchers suspended two copper particles in an aqueous solution between two electrodes. Under these circumstances, both particles become polarized, with electrons within the particles being drawn toward, or repelled by, the charge on the electrodes. For example, the electrode carrying an excess of electrons repels electrons within the particles and thereby induces a positive charge in the sides of the particles nearer to that electrode. The electrode with a deficiency of electrons attracts electrons in the particles and thereby induces a negative charge in the sides nearer this electrode. The charged particles are said to be bipolar because of the unequal distribution of electrical charges within them.

If, now, the electrical field between the two electrodes is increased, the distribution of charges within each particle becomes even more unequal. At some point, the electrical pressure within one particle becomes great enough that copper(II) ions are released from the particle. At the same time, decomposition of water begins to occur around the second copper particle. At some point, the concentration of copper(II) ions becomes great enough that ions begin to flow from one particle to the second particle. As they do so, they begin to deposit a very thin copper wire between the two particles. A connection has been made between the two particles simply by the application of a potential difference between the two electrodes.

Drexel researchers found that a variety of factors determined the growth of wires, the appearance of those wires, and the rate at which they grew. For example, with a field strength of less than 15 V cm^{-1} (volts per centimeter), no wire at all grew within a period of five minutes. As the electrical field strength was increased, however, wire growth was observed. With increasing field strength, the rate of growth also increased, up to a limit of about 35 V cm^{-1}. Field strengths greater than 35 V cm^{-1} did not further increase the rate at which wires grew.

The shape of the wire formed between two particles was also a function of electrical field strength. At relatively low field strengths, the wire tended to be more diffuse and "fractal" in its appearance. With increasing field strengths, the wire became narrower and more compact. Researchers report that their best estimate of the width of these wires is on the order of a few micrometers.

The appearance of wires formed by this method is also a function of the solution in which they were grown. For example, when a 2.5mM (millimolar) solution of $Cu(NO_3)_2$ is used (rather than water), the wires formed are wider and more diffuse than those grown in pure water or in more dilute solutions.

Another intriguing observation made during this series of experiments was that wire growth could easily be controlled by the proper positioning of surrounding electrodes. In a four-dot array, for example, wires could be grown between any two dots, depending on the orientation of the electrodes. When one electrode was placed outside the lower right dot

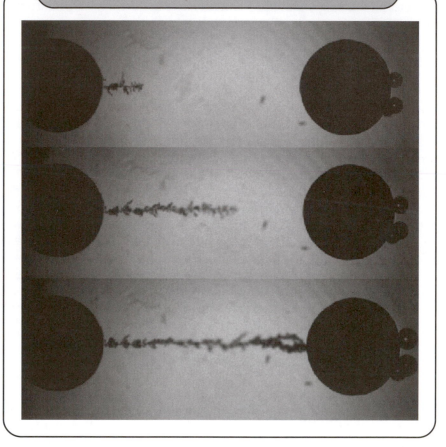

Fine copper wires are shown growing in water. *Courtesy of Jean-Claude Bradley.*

and the second electrode outside the upper left dot, a wire was generated diagonally between these two dots. When the electrodes were moved to the right and left of the four-dot square, however, wires were generated between the two lower dots and the two upper dots.

This observation suggests that it should be relatively easy to grow wires between any two given dots, leading to the possibility of wiring dots in a three-dimensional array, rather than in a purely two-dimensional array, as is now the case.

A Single-Molecule Rectifier

In 1974, Ari Avram, at the IBM Thomas J. Watson Research Center, and Mark A. Ratner, at Northwestern University, predicted that it might be possible to use single molecules as electrical rectifiers. A rectifier is a device that allows electrical current to flow with relative ease in one direction, but not in the opposite direction. Rectifiers are essential components of all electronic devices, and the development of a molecular rectifier would be an important step forward in solving problems of nanoelectronics.

Confirmation of the Avram-Ratner hypothesis did not appear for more than two decades. Then, in 1997, Robert M. Metzger at the University of Alabama at Tuscaloosa and his colleagues reported on the development of a molecule that can act as a rectifier.

The molecule they used in this experiment is hexadecylquinolinium tricyanoquinodimethanide, whose structure is shown below.

This molecule satisfies the general conditions of the Avram-Ratner hypothesis in that it contains three distinct parts: an electron donor (D), an electron acceptor (A), and a bridge (σ), all joined in the sequence D-σ-A. The characteristics of this molecule are such that it should be easy to convert D-σ-A to D$^+$-σ-A$^-$, but much more difficult to convert D-σ-A to D$^-$-σ-A$^+$.

In the hexadecylquinolinium tricyanoquinodimethanide molecule, the quinolinium segment—the double ring structure at the left of the molecule—acts as the electron donor part of the molecule. The tricyanoquinodimethanide segment, located at the right of the molecule, acts as the electron acceptor part of the molecule. Separating these two portions of the molecule is a cloud of electrons known as a π (pi) bond, which acts as a bridge across which electrons can travel.

This compound was tested as a rectifier in two ways. First, a layer only one molecule thick was sandwiched between aluminum electrodes. Next, a multilayer sandwich of molecules was inserted between the electrodes. When a potential of about 1 volt was applied across each of these systems, a flow of current was observed from "left" to "right," but not from "right" to "left." That observation would, of course, suggest that the molecule had behaved as a rectifier, allowing the flow of electrons in one direction only between the two electrodes.

Researchers concluded their report on this experiment by claiming that they had proved "that a monolayer of molecule 5 [the molecule described above] can rectify by intramolecular tunneling, and that monolayers and multilayers rectify both as macroscopic films and on a nanoscopic level" (Metzger, 10466).

Conductance through a Single Atom

Electronic engineers currently use very short wires or other small devices to connect the components of a computer chip. The smallest of those connecting devices is currently about 500 atoms in width. In order to make such connectors even smaller, one begins to think in terms of using even smaller numbers of atoms or, at the ultimate extreme, a single atom.

Not many years ago, the idea of measuring the conductivity across a single atom would have seemed absurd. The development of nanotechnology has made such experiments possible, however. Those experiments have not yet reached the point of practical applications—and probably will not do so for many years—but they have already been carried out in some laboratories.

The first of these experiments was reported in 1998 by a research team led by Elke Scheer at the University of Karlsruhe. In the Karlsruhe research, electrical currents were caused to pass across the diameter of single atoms of various metals, including gold, lead, aluminum, and niobium. The results of these experiments suggest that individual atoms are able to conduct electrical currents far better than currents measured on a macroscopic scale.

In order to make such measurements, the Karlsruhe team fabricated two pieces of aluminum film joined by a very thin bridge of aluminum. The fabrication and analysis of this structure was carried out by means of a scanning tunneling microscope built especially for this project. Using the microscope as a tweezers-like device, the team was able to break the aluminum bridge at its weakest point and insert a thin layer of metal. The inserted metal was then manipulated to reduce its width to a single atom.

As electricity was passed through this system, detectors recorded the amount of current flowing across the single atom. The results obtained confirmed the hypothesis that electron flow across a single atom is very different from electron flow across a piece of metal of macroscopic size. Single-atom flow obeys the principles of quantum mechanics, not those of classical physics. According to quantum mechanical principles, the resistance provided by a single atom is a fundamental constant equal to $h/2e^2$, which is equal to 12.9 kOhm.

Notice that the only factors determining this resistance are a universal constant, Planck's constant (h), and the electrical charge on the atom (e).

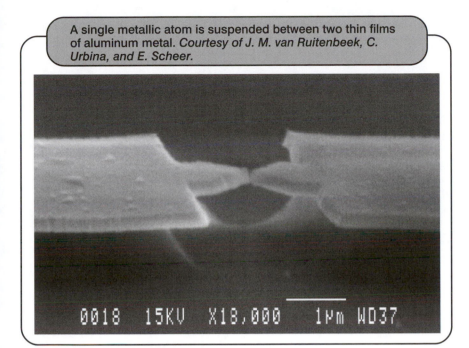

A single metallic atom is suspended between two thin films of aluminum metal. *Courtesy of J. M. van Ruitenbeek, C. Urbina, and E. Scheer.*

This fact would suggest that the only difference among various types of atoms used in this experiment would be the valence of the atom being studied. And this was precisely what the researchers found out.

They learned that a gold atom does have a resistance equal to that predicted by theory, 12.9 kOhm. By contrast, an atom of lead has a resistance only one-third as great (4.3 kOhm) because it has three valence electrons. On a macroscopic scale, by comparison, gold is 10 times more conductive than lead.

The magnitude of the flow of electric current through individual atoms is truly astounding. The authors of this study point out that a wire of macroscopic size capable of carrying a comparable current would have to carry 400,000,000 amperes. For the purposes of comparison, the typical electric wire used for household appliances melts at currents approaching 20 amperes.

Molecular AND/OR Gate

For all of their apparent complexity, computer chips actually perform a relatively small number of fundamental operations. One of those operations is a simple set of additions that can yield discrete results. These operations are known as AND/OR operations. The device that carries out these operations is called an AND/OR gate.

An AND/OR gate is a device in a circuit that adds two incoming inputs to produce a single output. It can differentiate among the following possible logical choices:

Case 1: no input from circuit 1 + no input from circuit 2 = no output

Case 2: input from circuit 1 + no input from circuit 2 = no output

Case 3: no input from circuit 1 + input from circuit 2 = no output

Case 4: input from circuit 1 + input from circuit 2 = output

These operations can be expressed mathematically as follows:

Case 1: $0 + 0 = 0$ Case 3: $0 + 1 = 0$

Case 2: $1 + 0 = 0$ Case 4: $1 + 1 = 1$

In 1997, A. Prasanna de Silva and colleagues at Queen's University in Belfast reported on the discovery of a molecule that can be used as a molecular AND/OR gate. The molecule consists of three fundamental parts, as shown below. In this diagram, the symbol CE represents a compound known as a crown ether. A crown ether consists of a group of carbon and oxygen atoms arranged in a cage-like structure. The term crown suggests the physical appearance of this kind of molecule.

The crown ether in the molecule is then attached to an anthracene molecule (represented by Anth in the diagram). Anthracene is an organic molecule consisting of three benzene rings joined to each other.

Finally, a second organic unit known as an amine is joined to the anthracene molecule at a position opposite to that of the crown ether. The amine unit is represented by the symbol $R-NH_2$ in the diagram.

$$CE — Anth — R-NH_2$$

The central anthracene unit can be made to fluoresce (give off light), provided that one of two (or both) conditions are met. First, fluorescence occurs when the crown ether sequesters a sodium ion. Sequestration is the process by which an atom or ion becomes trapped in the middle of the crown ether. Second, fluorescence also occurs when a proton is added to the amino group attached to the anthracene unit

$$R-NH_2 + H^+ \rightarrow R-NH_3^+$$

However, the fluorescence produced in either of these situations is very low. It amounts to $0.0051F_F$ and $0.0090F_F$ (where F_F is a unit of brightness), respectively. By contrast, the fluorescence produced when both events occur at once is $0.22F_F$, an increase of almost 40 compared with that observed when either of the two events occurs independently.

The molecule is operating in these cases, then, as an AND/OR logic gate. If it receives no input at all from either end (no sodium atom or proton; Case 1 above), it produces no output, that is, it does not fluoresce. If it receives an input from only one end (either a sodium atom or a proton; Cases 2 and 3 above), it produces no output. But if it receives both a sodium atom and a proton (Case 4 above), it produces an output: it fluoresces.

The success of this experiment has led to de Silva's growing optimism that molecular systems will soon be available for actual mathematical computations. "Hopefully," he has said, "we will use this type of molecule in arithmetic systems before too long" (de Silva, 7892).

The Technology of Nanotechnology

Electronic devices of the future will consist of very small components, perhaps no larger than single atoms and molecules. A problem facing engineers, then, is how to pick up, move, arrange, and otherwise manipulate these components.

Some solutions to this problem do not seem dramatically different from those currently used in the manufacture of microcircuitry. For example, a team of researchers at the Delft University of Technology in the Netherlands and the Institute for Inorganic Chemistry in Essen, Germany, under the direction of Cees Dekker, announced in 1997 the fabrication of a device for isolating and studying nanoparticles consisting of only a few thousand atoms.

To produce this device, the Dekker team first built a tiny pair of electrodes out of silicon on which a thin layer of platinum was deposited. They then inserted a drop of colloidal palladium metal in the space between the two electrodes. A small current applied to the electrodes kept the palladium in position while the water evaporated. After evaporation, the palladium droplet was held in position, much as if it were being grasped by a tiny pair of tweezers (the electrodes).

An even tinier device for working with nanoscale devices employs specially modified nanotubes attached to the tip of a scanning tunneling microscope (STM). Expanding on an idea originally proposed by Richard E. Smalley, the discoverer of buckyballs, Charles Lieber and his colleagues at Harvard University developed a tool that recognizes individual biological molecules. This tool is based on a standard STM, to which a nanotube is attached at the tip. A nanotube is used because its dimensions are much smaller than those of the typical STM tip, and it has the strength and flexibility to probe into the smallest spaces on the surface of the material being studied.

Lieber's team found that they were able to attach specific chemical species at the end of the STM nanotube tip. These groups are able to "recognize" chemical species on the surface being studied. Recognition occurs both because of matching geometric shapes in the nanotube and surface molecules and because of complementarity in chemical structures between such molecules.

The applications of this technique appear to be nearly endless. For example, one could construct a "toolbox" containing a variety of different tips for studying a particular surface, such as the outer structure of a cell membrane. By using one tip after another from the toolbox, a researcher could identify all of the chemical entities present on the surface and map their locations. The tips could also be used to study chemical reactions of molecules and ions in the surface, thereby identifying their function. Conceivably, it might also be possible to pluck off specific chemical species from the surface and perform more detailed analyses on them. Lieber has even predicted that it might be possible at some time in the future to use this tiny nanoprobe to remove specific defective segments of a DNA molecule, such as a mutant gene, and replace them with correct copies of those segments.

References

"AND Now, Molecular Logic Gates," *Chemistry & Industry,* 15 September 1997: 711.

Bradley, Jean-Claude, et al., "Creating Electrical Contacts between Metal Particles Using Directed Electrochemical Growth," *Nature,* 18 September 1997: 268–70.

Dagani, Ron, "Isolated Metal Dots Wired Electrochemically," *Chemical & Engineering News,* 22 September 1997: 9–10.

de Silva, A. P., H. Q. Nimal Gunaratne, and Colin P. McCoy, "Molecular Photonic AND Logic Gates with Bright Fluorescence and 'Off-On' Digital Action," *Journal of the American Chemical Society,* 20 August 1997: 7891–92.

Gibbs, W. Wayt, "From Chips to Cubes," *Scientific American,* November 1997: 45.

Glassman, Mark, "The Teeny-Weeny Frontier," *New York Times,* 4 August 1998: 8D.

Metzger, Robert M., et al., "Unimolecular Electrical Rectification in Hexadecylquinolinium Tricyanoquinodimethanide," *Journal of the American Chemical Society,* 29 October 1997: 10455–66.

Peterson, I. "One-Way Molecules Channel Electric Current," *Science News,* 8 November 1997: 293.

Scheer, Elke, "The Signature of Chemical Valence in the Electrical Conduction through a Single-Atom Contact," *Nature,* 9 July 1998: 154–57.

Schoelkopf, R. J., et al., "The Radio Frequency Single-Electron Transistor (RF-SET): A Fast and Ultrasensitive Electrometer," *Science,* 22 May 1998: 1238–42.

Service, Robert E., "Computers: Making Single Electrons Computer," *Science,* 17 January 1997: 303–04.

———, "The Fastest Counter of the Smallest Beans," *Science,* 22 May 1998: 1193.

Stix, Gary, "Is the End in Sight?" *Scientific American,* February 1998: 36.

Wong, Stanislaus S., et al., "Covalently Functionalized Nanotubes as Nanometre-sized Probes in Chemistry and Biology," *Nature,* 2 July 1998: 52–55.

QUANTUM DOTS

An exciting new field of research at the nanoscale level involves the study of quantum dots. Quantum dots are very small conductive regions contained within a semiconductor material. They are surrounded by materials that do not conduct the flow of electrons well. The nature of this arrangement is such that electrons trapped within a quantum dot are restricted to a relatively small number of discrete quantum states. In this regard, they are similar to atoms (whose electrons are also confined to specific quantum levels) and have, therefore, sometimes been referred to as mega-atoms or artificial atoms. They have also been called nanocrystals because they have characteristic, regular, crystal-like properties similar to those of macroscopic crystals.

Quantum dots can exist as flat two-dimensional structures or as box-like three-dimensional structures. The dimensions of a typical quantum dot are on the order of 100 nanometers. The number of electrons contained within a dot can vary widely and, in principle, can be as small as one.

The study of quantum dots is of great interest on both a theoretical and a practical level. On a theoretical level, they are the source of new information about the behavior of matter because they tend to follow the principles of quantum mechanics rather than those of classical physics. From a practical standpoint, quantum dots provide a new mechanism by which the flow of electrical current through a semiconductor can be controlled.

The number of studies on the properties and behavior of quantum dots has increased significantly in the last few years. The study reported below is provided as an example of that kind of research.

Manipulating Quantum Dots

In 1997, researchers at the Universities of California at Los Angeles and Berkeley explored the behavior of quantum dots as the distance between those dots was changed. The quantum dots in this experiment were made of silver and had diameters of the order of 20–40 Å. Each dot was chemically altered, or "capped," by being surrounded with an alkyl thiol group. An alkyl thiol group is an organic grouping that contains one or more sulfur atoms. The purpose of the capping was to maintain the

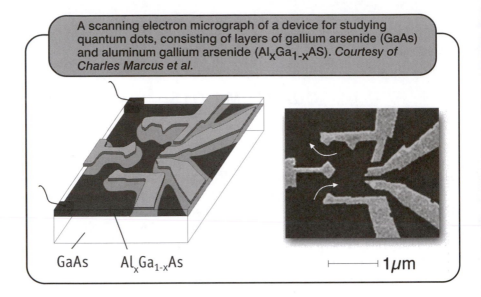

A scanning electron micrograph of a device for studying quantum dots, consisting of layers of gallium arsenide (GaAs) and aluminum gallium arsenide ($Al_xGa_{1-x}AS$). *Courtesy of Charles Marcus et al.*

GaAs $Al_xGa_{1-x}As$ ⊢——⊣ 1μm

integrity of each individual quantum dot and to permit the squeezing of dots without bringing them into physical contact with each other.

The dots were suspended in a monolayer on top of water at a distance from each other of about 12 Å. In this setting, each dot behaved as a single, discrete particle, a bit like a giant atom. The particle was a nonconductor, because all of the electrons within it were confined to the particle and were not free to carry an electric current. Standard tests on the quantum dots, such as a measure of their optical properties, indicated that they behaved as an insulator when arranged as described above.

The properties of the quantum dots began to change, however, as the distance between them was reduced. As the particles were forced closer and closer together, electron energy levels of adjacent particles began to overlap, and the assemblage of particles as a whole began to behave like a metal. That is, electrons no longer could be said to "belong" to any particular dot, but were free to travel from one dot to another. The assemblage had, in other words, taken on the properties of a metal and had become a conductor. The transition from insulator to conductor took place when the cores of the quantum dots were separated from each other by a distance of about 5 Å.

A second discovery of interest was that the transition from insulator to conductor was reversible. As soon as the forces pushing dots closer together were released, the dots once again moved away from each other, and the assemblage took on the properties of an insulator once again.

The results of this experiment suggest that it may be possible to invent new forms of matter with properties that can be specially designed for specific purposes. The quantum dots can be made of various metals, they can be capped with different materials, they can be dispersed with various distances between them, and they can be subjected to various external forces and influences. The specific combination of properties used in the development of any one nanocrystal assemblage will define the specific properties of the material finally produced.

References
Bains, Sunny, "Quantum Dots Get Organized" <http://www.spie.org/web/oer/november/nov96/quantum.html> accessed 30 April 1999.

Collier, C. P., et al., "Reversible Tuning of Silver Quantum Dot Monolayers through the Metal-Insulator Transition," *Science,* 26 September 1997: 1978–81.

Gammon, Daniel, "The Evidence of Small Things," *Science,* 10 April 1998: 225.

Gee, Henry, "Joining Quantum Dots" <http://helix.nature.com/nsu/981105/981105-3.html> accessed 30 April 1999.

Kastnere, M. A., "Artificial Atoms," *Physics Today,* January 1993: 24–31.

Landin, L., et al., "Optical Studies of Individual InAs Quantum Dots in GaAs: Few Particle Effects," *Science,* 10 April 1998: 262–64.

Peterson, Ivars, "Electrons in Boxes," *Science News,* 11 April 1998.

"Quantum Dots" <http://www.mitre.org/research/nanotech/quantum_dot.html> accessed 30 April 1999.

Reed, M. A., "Quantum Dots," *Scientific American,* January 1993: 118–23.

Wilson, Elizabeth, "Putting the Squeeze on Quantum Dots," *Chemical & Engineering News,* 29 September 1997: 9.

Further Reading
Jacak, Lucjan, et al., *Quantum Dots*, New York: Springer Verlag, 1998.

Turton, Richard, *The Quantum Dot: A Journey into the Future of Microelectronics*, Oxford: Oxford University Press, 1995.

MATERIALS

Scientists are constantly searching for new materials or new properties of well-known materials. The discovery of such materials makes possible, of course, new applications for products in everyday life. The following sections summarize some interesting discoveries made about the properties of materials in recent years.

A Thin Film with Reversible Optical Properties

Serendipity continually plays an important role in scientific discoveries. Serendipity refers to a discovery made by accident, often when a researcher is studying some unrelated phenomenon. In 1996, scientists at the Free University in Amsterdam, under the direction of Ronald P.

Griessen, reported on the serendipitous discovery of a material whose physical properties change dramatically when the pressure of hydrogen gas on the material is altered.

The Griessen team was actually studying an unrelated phenomenon at the time of the discovery—the superconductive behavior of hydrogen at low temperatures and high pressures. As Griessen tells the story, "One day, while we were exposing yttrium hydride to hydrogen in the laboratory, the sample just disappeared. It became invisible while the film was absorbing hydrogen" (Lipkin, 182). The yttrium hydride had been prepared in the form of a thin film about 500 nm thick. The film was coated with a thin layer of palladium metal about 5–20 nm thick to prevent the yttrium hydride from breaking down during the experiment. Palladium metal was used as a coating because palladium is porous to hydrogen gas.

With this arrangement, the yttrium hydride film was surrounded by hydrogen gas. When the pressure of the gas was increased, hydrogen passed through the palladium coating and reacted with the yttrium hydride film. A compound with the formula YH_x formed during this procedure, with the value of x depending on the amount of hydrogen passing through the palladium film. The pressure of the gas was increased gradually so the amount of hydrogen in the yttrium hydride also increased. The chemical structure of the yttrium hydride film gradually changed from YH to YH_2 to YH_3.

As these changes were occurring, researchers measured a number of physical properties of the rare earth hydride film, properties such as electrical resistivity, magnetoresistance, photoconductivity, Hall effect, optical transmission, magnetic field, and composition. The most striking change observed during changes in hydrogen pressure was optical. At low hydrogen pressures, the rare earth hydride film was totally opaque, as would be expected from a metallic surface. An object placed in front of the film produced a perfect reflection of itself, as would be seen in a silver mirror.

As the pressure of hydrogen increased, however, the film's reflectivity began to decrease. At a composition of $YH_{1.8}$, an object placed in front of the film was no longer perfectly reflected. Finally, at maximum hydrogen pressures, the film suddenly became totally transparent, losing one of its most obvious metallic properties. An object placed in front of the film at this point produced only a modest reflection that could be attributed entirely to the palladium coating. Researchers determined the composition of the film at this point to be $YH_{2.86}$. The turning point for this change appeared to take place rather quickly, as the composition of the yttrium hydride approached YH_2. At that point, optical transmission of the film increased dramatically in a matter of a few seconds.

The Griessen team pointed out in their report that their studies are only the beginning of a potentially more extensive project. Most rare earth elements have properties similar to those studied for yttrium, and it remains to be seen how other members of the family will behave when exposed to conditions similar to those reported in this study. Potential practical applications are thought to be still some time off, but they may

The yttrium dihydride film acts as a switchable mirror—upon exposure to hydrogen, it becomes transparent. *Courtesy of R. Griessen, Vrije Universiteit, Amsterdam.*

include uses in such devices as solar cells, optical sensors, and windows for homes and office buildings.

References

Huiberts, J. N., et al., "Yttrium and Lanthanum Hydride Films with Switchable Optical Properties," *Nature*, 21 March 1996: 231–33.

Lipkin, R., "Thin-Film Mirror Changes into a Window," *Science News*, 23 March 1996: 182.

Anisotropic Chalcogenide Glasses

Glassy materials are generally thought to be isotropic, that is, they transmit light equally in all directions. Scientists have known for some time, however, that one type of glass is anisotropic, that is, it tends to transmit light more readily in one direction than in another. These glasses are the chalcogenide glasses. The chalcogenides are elements in column 16 of the periodic table and include oxygen, sulfur, selenium, and tellurium.

In 1997, a group of researchers at the University of Cambridge in England reported on an interesting experiment in which chalcogenide glasses were made to change their shape when polarized light was shined on them. In this experiment, the Cambridge researchers constructed a microlever from a thin strip of silicon nitride 200 μm long, 20 μm wide and 0.6 μm thick. They deposited a thin film (250 nm thick) of chalcogenide glass on top of the microlever. The glass was made of an amorphous material with the general formula AsSe.

To study the effects of light on the chalcogenide film, they shined a beam of polarized light at the microlever along its length and across its width. They discovered that the microlever changed position upward or downward by about 1 nm, depending on the direction from which the polarized light struck the microlever. If one imagines the original chalcogenide film as having a cylindrical shape at the beginning of the experiment, the circular cross-section of that cylinder was "squashed" in either one direction or the other by the polarized light.

This experiment is important for two quite different reasons. In the first place, it provides researchers with useful information about the molecular structure of chalcogenide glasses that had not been available from other studies. It appears that the light-dependent shape changes produced in the material result from changes in the AsSe dipoles that make up the molecular structure of the chalcogenide glass. These changes involve relatively dramatic reorientations of these dipoles from random orientations to parallel, ordered orientations in one or another direction, depending on the polarization of light used.

The experiment is also important because of its potential application in meeting future technological needs at the nanoscale level. Currently, there are relatively few methods by which structures at the nanoscale dimension can be manipulated, as would be necessary in the construction of nanoscale electronic devices. The Cambridge discovery makes it clear that nanoscale components can be made to change shape relatively easily simply by shining polarized light on them. One could imagine using this technique, for example, for the manipulation of wires and switches in order to make contacts in a nanoscale transistor.

References

Krecmer, P., et al., "Reversible Nanocontraction and Dilatation in a Solid Induced by Polarized Light," *Science,* 19 September 1997: 1799–1802.

Tanaka, Kazunobu, "Light-Induced Anisotropy in Amorphous Chalcogenides," *Science,* 19 September 1997: 1786–87.

Wu, C., "Glass Film Yields to a Light Touch," *Science News,* 20 September 1997: 183.

OPTICS

The studies of the properties of light waves—optics—is one of the oldest and newest fields of research in physics. At one time, optical studies concentrated on the way light is reflected and refracted through various objects. Such studies resulted in the invention of a number of invaluable tools, such as microscopes, telescopes, and spectroscopes.

Today, the challenge in optics is somewhat different. An important focus of research in physics today is on smaller and smaller structures: nanotubes, quantum dots, and other objects approaching the size of atoms and molecules. In order to study and work with such objects, more precise tools are needed than those that operate with electrical current. A flow of electrons is often too crude a tool to use on nanoscale devices. What physicists would prefer to do is to use light, which has a much smaller wavelength than a flow of electrons. Much of the research in optics today, then, is in an effort to learn more about the properties of light waves and other forms of electromagnetic energy and to find ways of putting those properties to use in practical devices. The studies below are typical of that kind of research.

A Photonic Crystal

Semiconductors are the key to the modern electronics industry. One reason they are so valuable in electronic devices is that their structure contains very specific regions in which electrons may and may not exist. The "forbidden" regions in a semiconductor are known as electronic bandgaps. Engineers can control the structure of electron bandgaps by "doping" semiconductors with a variety of materials (adding impurities of one kind or another, such as arsenic and phosphorus). The manipulation of electronic bandgaps has made possible many of the special functions that electronic devices are capable of carrying out.

As engineers explore ways of making electronic devices more efficient, they hope to find ways of using light rather than electric currents to power the devices. The speed at which light moves is many, many times greater than the speed of an electric current. An electronic device that operates with light rather than electricity will, then, be far more efficient than existing devices.

The basic problem for engineers, then, has been to find a way of capturing photons, the particles of which light is made, and manipulating them the way electrons are captured and manipulated in semiconductors. In 1997, physicists at the Massachusetts Institute of Technology (MIT) reported on the development of a method for achieving this kind of control of photons.

The MIT team accomplished this task by means of a technique that is simple in concept, although very difficult in execution. They produced a photonic bandgap comparable to an electronic bandgap by boring tiny holes in a thin strip of silicon. The holes in the silicon were about one micron (one micrometer, or one-millionth of a meter) apart. This spacing is roughly comparable to the wavelength of visible light.

The MIT invention had one additional property. The space between two of the holes in the middle of the strip was somewhat larger than that between the other holes. This wider spacing acts as a barrier that limits the wavelength of radiation that is able to pass through the strip. In this configuration, the strip is able to trap infra-

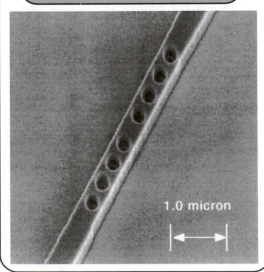

A "photonic bandgap" is created by drilling this tiny strip of silicon with accurately spaced holes that block certain wavelengths of radiation. *Courtesy of John D. Joannopoulos, MIT.*

1.0 micron

red radiation of the type currently used for communication systems based on fiber optics.

The key to the success of the MIT invention was finding a mechanism for drilling tiny holes spaced so close to each other on the silicon strip. To do so, they used X-ray lithography, a technique in which beams of focused X rays are used to make very precise cuts of very small size in a material.

The MIT photonic device has great promise as a substitute for, and an improvement on, semiconductor materials. As with a semiconductor, the properties of a photonic device can be altered in a variety of ways. Rather than adding various dopants, the photonic devices can be changed by altering the number, location, and spacing of holes drilled in the silicon strip. Among the many applications that the MIT teams sees for its invention is the production of very small portable light sources, such as lasers. The photonic device would act as the optical cavity in which light is reflected repeatedly in a laser.

References

Foresi, J. S., "Photonic-Bandgap Microcavities in Optical Waveguides," *Nature*, 13 November 1997: 143–45.

Hayasi, Alden M., "Let There Be No Light," *Scientific American*, February 1998: 32.

Taubes, Gary, "Photonic Crystal Made to Work at an Optical Wavelength," *Science*, 5 December 1997: 1709–10.

Wu, C., "Holey Device Traps Light for Lasers, Filters," *Science News*, 15 November 1997.

X Rays from Visible Light

Lasers have revolutionized many aspects of the modern world. They have become invaluable tools not only for scientific researchers studying nearly every aspect of the living and nonliving world, but also for an almost endless variety of devices in everyday life, from CD players to surgical instruments.

For a number of years, scientists have been searching for ways of extending the principle of laser light production to the production of high intensity, coherent X rays. An X-ray laser would have many new applications because of its ability to penetrate through materials that are currently opaque to the visible light from optical lasers.

In 1998, physicists at the University of Michigan announced the development of the first device that converts visible light to X rays for use in an X-ray laser-type device. It consists of a conventional optical laser attached to a glass tube three centimeters in length that acts as a waveguide. As light from the laser passes through gas in the waveguide, it occasionally knocks electrons loose from their nuclei. When those electrons return to their ground state in the atom, they release a very large pulse of energy with frequency in the X-ray region. This process is known as harmonic conversion.

The problem with the system as described is that X rays produced by harmonic conversion fly though the tube at high speed, while photons from the laser beam are slowed down by the gas within the tube. As a consequence, X rays produced earlier in the process get far ahead of those produced later, when the laser beam is still in the back part of the tube. The X rays produced are intense, but they are not coherent.

To solve this problem, the Michigan researchers took two steps. First, they designed the glass waveguide tube so that the photons passing through it would not slow down as much when they pass through the gas. In fact, the term waveguide refers to a system in which the shape of a wave can be very precisely controlled.

The researchers also controlled the speed of photons in the waveguide by altering the pressure of the gas in the tube. They found that the "magic pressure" they needed to produce a coherent X-ray beam was 30 torr. At that pressure, a coherent X-ray beam shot out of the tube with "surprising intensity," according to Andy Rundquist, a member of the Michigan team. "At two o'clock in the morning, I was yelling and screaming in the lab," he was quoted as saying (Kestenbaum, 1348).

The Michigan success did not meet all of the expectations physicists have for an X-ray laser. For example, the X rays produced in the Michigan device are "soft" X rays, those with relatively low frequency. Most researchers are hoping to develop a laser with more powerful X rays that can "see through" materials such as water. It is lasers of this kind that will probably find a wider variety of applications than the Michigan model. Nonetheless, the Michigan experiment has broken through an important barrier in the design of X-ray lasers, and it points the direction for future research in this area.

References
Kestenbaum, David, "Transmuting Light into X-rays," *Science,* 29 May 1998: 1348.
Rundquist, Andy, et al., "Phase-Matched Generation of Coherent Soft X-rays, *Science,* 29 May 1998: 1412–15.

Further Reading
National Research Council, *Harnessing Light: Optical Science and Engineering for the 21st Century*, Washington, DC: National Academy Press, 1998.

CHAPTER THREE
Unsolved Problems

INTRODUCTION

From time to time, scholars appear who announce the end of scientific progress. Convinced that all of the great discoveries that *can* be made *have* been made, they argue that science of the future will be only a dim afterglow of all that has gone before.

Such naysayers have always been wrong. Researchers continue to find new and exciting problems on which to focus their attention, often with startling new breakthroughs. The development of nanotechnology, described in Chapters 1 and 2, is an illustration of how this process continues today.

What problems and challenges, then, do physicists still face? The answer to that question is multifaceted. Dozens of intriguing questions about the natural world remain. In this chapter, we review a handful of the most significant, most exciting, and most pregnant of those questions.

DARK MATTER

One of the most frustrating puzzles in astrophysics goes back more than half a century. That puzzle involves the mystery of dark matter. *Dark matter* is a term that refers to matter that astronomers believe exists in the universe, but which, so far, has not been actually detected.

Historical Background

The dark matter puzzle originates with the research of the Swiss astronomer Fritz Zwicky (1898–1974). In the early 1930s, Zwicky was concerned with measurements of the mass of galaxies. Such mass cannot be measured directly, of course, but can be inferred by two different but reliable methods. The first of these methods uses the luminosity (brightness) of a galaxy. The relationship between the amount of matter in a star or a galaxy and the amount of light it produces was well known in Zwicky's time. He used this relationship to calculate the mass of a number of distant galaxies.

A second method for determining astronomical masses is by measuring the movement of objects. The speed with which the Earth moves around the Sun, for example, is dependent in part on the masses of the Earth and the Sun. Similarly, the mass of a rotating galaxy can be inferred from the speed with which the galaxy rotates.

To Zwicky's great surprise, the masses he calculated for a number of distant galaxies differed, depending on which of these two methods he used. In some cases, he found that the mass determined on the basis of velocity was 400 times greater than that determined on the basis of luminosity.

Zwicky drew a simple, but astonishing, conclusion from his studies. A large portion of the mass in the galaxies he was studying, he said, must be in the form of nonluminous matter, that is, matter that gives off no light. That "missing matter" was later given the name *dark matter*.

Zwicky's results were largely ignored by his colleagues. Many thought that he had made fundamental errors in his research. In any case, they knew of no explanation for the existence of dark matter in the universe.

Current Beliefs

It was not until the 1970s that astrophysicists began to think seriously about dark matter. One factor that led to this reevaluation of Zwicky's work was the increasing amount of evidence in support of the existence of dark matter. Measurements at every astronomical level—from those made in the neighborhood of our own solar system to other regions of our galaxies to more distant galaxies to the largest and most distant clusters of galaxies—all revealed the problem of missing matter. The amount of missing matter ranges from a quantity about equal to that of known matter in a region to a quantity hundreds of times greater than that of known (luminous) matter. Some astronomers now believe that 99 percent of all the matter in the universe may be in the form of dark matter.

How could one account for such large amounts of unobservable matter in the universe? Over the past few decades, astrophysicists have been hypothesizing answers to that question and devising experiments to test their hypotheses. Currently the candidates for dark matter tend to be classified into one of two categories: MACHOs or WIMPs.

MACHOs is an abbreviation for MAssive Compact Halo Objects and refers to objects that were first hypothesized as present in the halo of matter that surrounds galaxies. The term is now used in a more general way to refer to any form of cold matter made of baryonic matter. *Baryonic* matter is matter made of baryons, such as protons, neutrons, and electrons. It is matter that is common and with which scientists are very familiar.

Among the candidates for MACHOs in the universe are objects such as small, dim stars, such as brown dwarfs and black holes. Brown dwarfs are stars that "never made it," that is, they are too small for nuclear reactions ever to have started within them. They never became "real" stars that give off heat and light. They may be abundant in galaxies throughout the universe, but they would be difficult to find since they give off very little radiation.

Black holes, by contrast, are the remnants of "real" stars that have reached the final stage of their lives. They form when nuclear reactions within a star come to an end, and the star's gravitational field pulls all of its matter together in a single point. That single point represents an enormous concentration of mass that does not, however, give off any light.

WIMPs are the opposite of MACHOs in most respects. In the first place, they are made of non-baryonic matter, objects such as neutrinos, which are not as well understood as protons, neutrons, and electrons. The term WIMP itself stands for Weakly Interacting Massive Particle.

WIMPs fall into two categories or "flavors": cold dark matter and hot dark matter. Cold dark matter is thought to consist of relatively massive particles that move quite slowly and give off neither heat nor light. No such particles have ever been observed, but theorists have predicted a variety of particles that might fit this description. One example is the photino. The *photino* is an "antiparticle form" of the photon, predicted by the modern theory of supersymmetry. The photino has been hypothesized as having 10 to 100 times the mass of a proton.

By contrast, hot dark matter consists of particles with very little mass that move at close to the speed of light. Such particles tend to pass through "normal" (baryonic) matter without undergoing any interactions. Since no interactions occur, it is difficult for scientists to detect their presence.

One of the most attractive features of the hot WIMP hypothesis is that particles fitting this description have already been discovered. One promising candidate for a hot WIMP, for example, is the neutrino. When first hypothesized by Wolfgang Pauli in 1930, the neutrino was thought to have no mass at all or, at the most, a very small mass. Until the late 1990s, physicists were still in disagreement as to whether the neutrino has mass. Then, in 1997, the question was resolved experimentally: Studies showed that some types of neutrinos may, in fact, have an extremely small mass. That mass is much less than that of an electron, the least massive of the baryons.

How could a particle with as little mass as the neutrino help solve the problem of dark matter? The answer to that question lies not in the mass of a single neutrino, but in the total number of neutrinos present in the universe. Untold numbers of neutrinos were created during the Big Bang, and most of them are still traveling unhindered through space. The vast number of these particles makes up many times over for the very small mass of any single one of them.

Looking for MACHOs and WIMPs

The methods scientists use in looking for MACHOs and WIMPs are, as one might expect, very different. One of the most promising techniques for finding MACHOs is called *gravitational lensing*. Gravitational lensing makes use of scientific principles that are very similar to those that explain the focusing of light by a glass or plastic lens. As light passes through such a lens, it is bent. An observer on one side of the lens sees an object on the other side of the lens as an image that is magnified (or reduced, depending on the kind of lens).

Einstein's General Theory of Relativity says that massive bodies warp the space-time continuum around them. When light travels close to such bodies, then, its path is bent. The gravitational field of the body acts like a massive lens to distort the image of any objects behind it (when compared with an observer).

Gravitational lensing has begun to produce evidence for the existence of certain types of MACHOs. In 1995, for example, a team of American and Japanese scientists reported on the discovery of a black hole with a mass equal to that of 36 million Suns. Other researchers have discovered a number of MACHOs in the halo surrounding the Milky Way.

The search for WIMPs takes a very different direction. The challenge is to find a way to detect one of the very few occasions in which a neutrino or other WIMP particle interacts with matter. As an example, one project involves the use of crystals that have been cooled close to absolute zero. At this temperature, the motion of atoms and ions in the crystal has nearly ceased. The interaction of a WIMP particle with an atom or ion in such a crystal will produce an increased level of activity that is relatively easy to measure.

Cosmological Consequences of Dark Matter

The puzzle of dark matter has profound implications for the study of cosmology. *Cosmology* deals with the origin of the universe and its ultimate fate.

The first way in which dark matter has cosmological significance is with regard to the "clumping" of matter that occurred after the Big Bang. Most scientists today believe in the Big Bang but are troubled by at least one important discrepancy. What accounts for the fact that matter formed in the Big Bang soon began to clump together to form atoms, stars, galaxies, and other collections of matter? No explanation exists to show why matter clumped together rather than simply flying apart.

Some astrophysicists believe that WIMPs may provide the answer to that puzzle. It may be that neutrinos or other very light particles exerted sufficient gravitational force on other forms of matter to bring them together into the clumps that we now see today.

Dark matter is also important in understanding what the ultimate fate of the universe is to be. According to current cosmological theories, the universe can be described in one of three ways, defined by its density, designated by the Greek letter omega (ω). If omega is less than 1, the universe is expected to expand forever. If omega is greater than 1, the universe will eventually stop expanding, as it is now doing, and will eventually begin to collapse. If omega is exactly 1, the universe will remain essentially in its current condition and will neither expand nor collapse in the long-term future.

It is currently possible to begin estimating the ultimate fate of the universe based on the mass of observable objects in the universe. These measurements suggest that the universe does not contain enough mass for it to hold together forever, that is, the measured value of omega is considerably less than 1.

But what if very large amounts of dark matter do exist? In that case, the real value of omega might be much greater than 1, and the universe would someday stop its expansion and begin to contract, ending in what some physicists have called the Big Crunch. Of course, none of us will be around to observe these changes, but knowing what lies in store for the universe is one of the most fundamental questions in all of science.

References
Cole, K. C., "Halo around Milky Way Is Reported," *Los Angeles Times*, 5 November 1997: A1.

Griest, Kim, "The Search for the Dark Matter: WIMPs and MACHOs," *Annals of the New York Academy of Sciences*, 15 June 1993: 390–407.

Miller, Chris, "Cosmic Hide and Seek: The Search for the Missing Mass," <http://www.eclipse.net/~cmmiller/DM/> accessed 31 May 1999.

NASA, "Imagine the Universe!" <http://imagine.gsfc.nasa.gov/docs/homepage.html> accessed 30 April 1999.

Primack, Joel R., "A Little Hot Dark Matter Matters," *Science*, 29 May 1998: 1398–1400.

Silk, Joe, "Dark Matter," <http://cfpa.berkeley.edu/darkmat/essay.html> accessed 30 April 1999.

Trefil, James, "Dark Matter," *Smithsonian*, June 1993: 27–35.

Wilford, John Noble, "Physicists Step Up Exotic Search for the Universe's Missing Mass," *New York Times*, 26 May 1992: C1.

THE HIGGS BOSON

It is sometimes easy to take the most fundamental ideas of all for granted. Mass is an example. Almost everyone understands the concept of mass, whether it be at the simplest experiential level or at a more advanced and abstract level. That is, we know that matter "weighs" something. But the interesting question one might ask about mass is, "Where does it come from?"

We might expect that the answer to that question will be a simple one. It must be that the laws of physics require or demand the existence of mass. And from one perspective, that is true. Newton's laws of motion, for example, certainly incorporate and depend on the concept of mass. But those laws are really nothing other than generalizations of observations made about the way matter behaves. They say that some objects are observed to have more or less mass than others. But they do not say where that mass originates.

In fact, one of the intriguing points about modern physical theories is that there is nothing in those theories that accounts for the property we know of as mass. The equations of quantum chromodynamics (QCD), for example, fully account for the behavior of fundamental particles without imputing to them the property of mass.

Thus, one of the challenges facing physicists today is to find a way of accounting for one of the best known and most common of all properties of matter: mass.

One of the most popular answers to this question was first proposed in the 1960s by the Scottish physicist Peter Higgs. Higgs suggested the existence of a field—something like an electromagnetic field—that permeates all of space. As particles travel through space, they gain mass from the Higgs field.

One might compare this phenomenon to the effort of a waterbug to swim across the top of a jar of honey. The viscous nature of the honey impedes and slows down the waterbug's movement. The waterbug seems to gain mass as it travels across the honey.

The Higgs field is a bit more complex than this analogy suggests, however. For one thing, some particles gain more mass than others as they move through the field; that is, the field appears to have properties that make it easier to travel in one direction than another. The only problem with this description is that the Higgs field has no properties at all. In fact, it is distinguishable only because of the effect it has on particles moving through it.

Finding such a field obviously presents some serious problems. Scientists are, therefore, pursuing this search from another angle. According to quantum mechanics, every field has a corresponding particle, and every particle has a corresponding field. In fact, there is no distinction between particles and fields in quantum mechanics. A photon traveling through space, for example, sometimes behaves like a massless particle and sometimes like a wave.

The particle that corresponds to the Higgs field is known as the *Higgs particle* or *Higgs boson*. The Higgs boson is a most unusual particle in that it has only one property: mass. It lacks all of the other characteristic properties of particles, such as spin and charge. Its mass has been estimated to be about that of an atom, anywhere from that of an iron atom to three times that of a uranium atom.

A particle that massive can exist only at very high energies. There are no particle accelerators currently in operation that can produce the amounts of energy needed to produce a Higgs boson. The Superconducting Super Collider (SSC) planned, designed, and approved in the United States in the 1980s would have attained these levels. Now, the conditions needed to produce the Higgs boson will not be available until the Large Hadron Collider (LHC) at the Centre Européen pour la Recherche Nucléaire (CERN) comes on line in 2005.

The Higgs boson holds promise for theoretical physics in other respects also. If it exists, it could be a key to the unification of all known physical forces. As an example, physicists now have a theory that explains how two of the four major forces (gravitation, the strong force, the weak force, and electromagnetic force) are related to each other. The Electroweak Theory, developed by Steven Weinberg, Abdus Salam, and Sheldon Glashow (independently) in 1967, shows how the weak force and electromagnetism are two manifestations of a single force, the electroweak force.

The problem arises with the way in which the two manifestations of the electroweak force are expressed. The weak force is carried by two massive particles, the W and Z bosons, while electromagnetism is transmitted by a massless force, photons. How does it happen that these particles, both carriers of the electroweak force, could have such radically different masses? No one yet knows how, but many theorists hope that the Higgs boson might be involved in the answer to that question.

The search for a Higgs field, the Higgs boson, and perhaps groups of Higgs bosons is one of the most exciting and most promising frontiers of theoretical physics today.

References

Physics World. Five user-friendly explanations of the Higgs boson were printed in the September 1993 issue of *Physics World*. They can be read at <http://hepwww.ph.qmw.ac.uk/epp/higgs1.html>, <.../higgs2.html>, <.../higgs3.html>, etc.; accessed 30 April 1999.

Quigg, Chris, "Elementary Particles and Forces," *Scientific American*, April 1985: 84–95.

Quigg, Chris, and Roy F. Schwitters, "Elementary Particle Physics and the Superconducting Super Collider," *Science*, 28 March 1986: 1522–27.

Veltman, Martinus J. G., "The Higgs Boson," *Scientific American*, November 1986: 76–84.

PARTICLE ACCELERATORS

Particle accelerators are somewhat peculiar machines. To begin with, they are essential to the work of particle physicists because they provide an energy level that cannot be achieved by any other means on Earth, natural or synthetic. At the extremely high energy levels of particle accelerators, physicists gain insight on aspects of the nature of matter that remain forever hidden at the "normal" energy levels that are experienced in everyday life.

For that reason, progress in particle physics is impossible without better and better accelerators, which results in a second characteristic of these machines: They are always out of date. No sooner does construction begin on the newest, most powerful particle accelerator in the world than physicists begin to think about, design, and look for funds for its successor—almost by definition a more powerful machine than the one that has not yet begun operation.

On the other hand, particle accelerators have relatively few practical applications. When physicists go to national governments to ask for the millions or billions of dollars needed to build one, they can seldom promise cures for diseases, new and better building materials, or other improvements in human life. The primary (although not exclusive) function of these machines is to reveal more detailed information about the nature of matter.

The SSC and Beyond

In recent decades, the most striking example of the problems posed by accelerator construction is that of the Superconducting Super Collider (SSC). This powerful accelerator was more than a decade in the planning before it was approved for construction by the U.S. Congress in 1987. Only after it had been under construction for about six years did Congress, scientists, and the general public realize how great the actual cost would be. In 1994, when the estimated construction costs finally reached $12 billion—at least triple the amount originally estimated—Congress withdrew federal funding and canceled the project.

At that point, U.S. particle physicists were confronted with a serious problem: To continue their research, they might have to travel to other parts of the world where powerful accelerators were either in existence or were being planned. In most cases, that meant a trip to the Centre Europeén pour la Recherche Nucléaire (CERN) in Geneva, Switzerland. CERN not only housed some of the most advanced accelerator technology in the world, but it was also about to begin construction on an even more powerful machine, the Large Hadron Collider (LHC), due to begin operation in the early twenty-first century.

Even before the SSC was canceled, however, U.S. physicists were thinking about and planning for other kinds of particle accelerators. As the twentieth century drew to a close, most particle accelerators could be classified into one of two general categories: hadron colliders or lepton linear accelerators (linacs). Hadrons are subatomic particles with relatively large mass, such as protons. Leptons are subatomic particles with relatively little mass, such as electrons.

In hadron colliders, two beams of hadrons circulate in opposite directions until they reach speeds close to the speed of light. They are then diverted from their circular paths and caused to collide with each other. The Super Proton Synchrotron (SPS) at CERN, for example, accelerates protons and antiprotons in opposite directions before causing them to collide with each other. The energies released in such collisions reach into the low teraelectron (billion electron) volts (TeV).

Linacs, by contrast, are accelerators that cause lightweight particles such as electrons to travel over long distances in straight lines. Thus the term *linear accelerator,* or *linac.* These particles may then be caused to collide with other particles traveling from the opposite direction. The SLAC at Stanford University, for example, is designed to collide electrons with their antiparticle twins, positrons, with the release of a few billion electron volts of energy.

Future Particle Accelerators

Traditional particle accelerators, like the SPS and SLAC, have produced enormous volumes of invaluable information about the nature of matter. However, both types of accelerators also have inherent problems that limit their future usefulness. For example, hadron colliders like the SPS produce very large numbers of different kinds of particles. Sorting out information that is needed from "junk" particles is often very difficult. For example, when protons collide with antiprotons, many different kinds of quarks, gluons, and other fundamental particles are produced. Finding just the right particle in which one is interested—such as a Higgs boson—can involve a frustrating search for a "needle" in a "haystack" of less interesting particles.

In the case of electron accelerators, the problems are somewhat different. Fewer "junk" particles are formed, but the analytical value of the electron beam is itself limited by unavoidable energy losses through processes such as synchrotron radiation. Although synchrotron radiation has turned out to have useful by-product effects, as in the treatment of cancer, in and of itself it is a hindrance rather than an aid to understanding the composition of matter.

Partially in response to issues such as these, particle physicists have looked beyond traditional accelerators to visualize other types of machines with even greater potential than hadron colliders or electron linacs. Two such machines are muon colliders and gamma-ray colliders. Both machines pose difficult technical challenges, however, which will need to be solved before they can enter the stages of construction and use.

Muon Colliders

In the case of muon colliders, for example, the most obvious challenge is particle lifetime. Muons have a lifetime of about 2.2 microseconds. That lifetime can be increased substantially by accelerating them to nearly the speed of light. As a result of relativistic effects (as objects travel more rapidly, their time frame slows down), the lifetime of a muon can be stretched to as much as a few milliseconds. But that period of time is still very short in terms of the events that must take place in a particle accelerator. They must be created, captured, accelerated, and collided with other particles in a period of time far less than that with which physicists are used to working when using an accelerator.

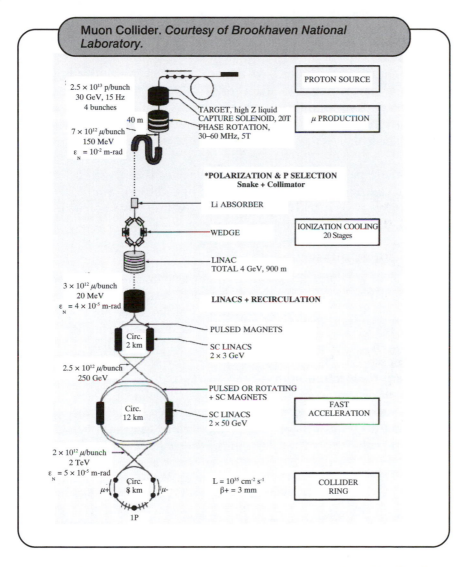

Muon Collider. *Courtesy of Brookhaven National Laboratory.*

In one design of a muon accelerator, muons are produced by the decay of pions, particles produced abundantly during proton-antiproton collisions that have a somewhat longer lifetime than muons. The muons formed in this process are heavier cousins of the electron with a mass about 200 times that of an ordinary electron. Also, like the electron, muons exist in two complementary forms, muons and antimuons.

The muons produced by pion decay are fed rapidly into a collider ring in which muons travel in one direction and antimuons travel in the opposite direction. In a typical collider ring, particles travel around the ring millions of times before they are forced to collide. By contrast, muons will be able to make no more than about 1,000 orbits before a collision takes place.

Muon-antimuon collisions have the advantage over electron-positron collisions of developing a greater energy yield and, hence, the possible production of particles with even greater masses than have been observed before. Such collisions will be possible because of the absence of synchrotron radiation that limits the energy developed by electron-positron collisions.

One of the particles that may be produced in an energetic muon-antimuon collision is the hypothesized Higgs boson. Should such a particle be created, it will be much easier to find than if it had been produced in a more typical hadron collider.

Gamma-Ray Collider

The design of a gamma-ray collider would also be somewhat different from that of traditional particle accelerators. Perhaps the simplest approach would be to make use of an existing linear accelerator to produce a beam of high-energy electrons. When this beam of electrons is focused on a beam of high-energy photons, as can be produced by a laser, a process known as back-scattering occurs. In *back-scattering*, the collision of an energetic electron with a high-energy photon can result in the formation of a gamma ray.

The gamma-ray collider must be designed, then, so that two gamma rays produced by this process can be accelerated at each other at high velocity and in a narrow beam. When the two gamma rays collide, there is the possibility for the creation of a new particle of large mass. Recall that because of the equivalency of mass and energy ($E = mc^2$), the more energy available in a particle beam, the more massive a newly created particle can be. Again, the hope is that a gamma-gamma collision could yield enough energy to result in the formation of a Higgs boson.

Beyond the search for the Higgs boson, particle physicists have imagined many other products of muon-antimuon and gamma-gamma collisions. For example, some physicists think that quarks and leptons may not be fundamental particles, but, like protons and neutrons, may themselves consist of even more fundamental particles. In order to explore that level of matter, however, even higher levels of energy than those now available with existing machines must be produced. Muon-antimuon and gamma-gamma machines may be just the accelerators needed for this research.

References

Durrani, Matin, "Dreams of a Muon Collider," *Physics World*, May 1997: 8.

Hellemans, Alexander, "Physicists Dream of a Muon Shot," *Science*, 9 January 1998: 169–70.

Kestenbaum, David, "Reports Call for New Super-Accelerator," *Science*, 27 February 1998: 1296–97.

"Muon-Muon Collider Collaboration Web Page" <http://www.cap.bnl.gov/mumu/mu_home_page.html> accessed 30 April 1999.

Sessler, Andrew M., "Gamma-Ray Colliders and Muon Colliders," *Physics Today*, March 1998: 48–53.

"U.S. Signs On to Collider Project," *Science*, 12 December 1997: 1871.

STRING THEORY AND M-THEORY

Historically, one of the most basic challenges in all of science has been to find out what is truly fundamental in nature. Even before modern science was born, ancient Greek philosophers were talking about the fundamental "elements" of nature, such as earth, air, fire, and water or mercury and sulfur.

During the past 200 years, scientists decided first that atoms; then, that protons, neutrons, and electrons; and eventually, that quarks and leptons were the fundamental components of matter. Photons, certain types of bosons, and the never-seen graviton have been designated as the comparable particles of energy.

The nature of these fundamental particles and their relationship to each other is currently described in a theory known as the Standard Model (see the table on page 3). The Standard Model has been the most complete and useful "guide" to the ultimate composition of matter and energy for nearly three decades. It has guided and explained much of the theoretical speculation and experimental research in particle physics for many years.

Remaining Problems

Yet, scientists are not totally satisfied with the Standard Model. Indeed, many physicists believe that, as useful as this theory has been, it provides no more than a hint about the true underlying nature of matter.

At least two basic problems remain concerning which the Standard Model provides little or no help. The first problem is whether the particles that make up the Standard Model itself—quarks and leptons—are truly fundamental or whether they, like protons and neutrons, are made up of even simpler particles. Quarks, leptons, and certain force-carrying particles currently form the basis of the Standard Model, at least partly because the particle accelerators currently available to study the composition of matter are not energetic enough to generate even more basic particles even if they did exist. But what particles and what combinations of particles might have existed at the even higher energy levels that we know existed at least once in the history of the Universe, at the time of the Big Bang?

A second problem that troubles physicists is that the two most basic physical theories in use today—quantum mechanics and general relativity—provide very different, indeed, contradictory, explanations of the nature of gravitational force. And since gravity really does exist and is well understood, some explanation must be developed that will show how quantum mechanics and general relativity can both be correct at the same time, even though they provide different views of gravity.

An even more overarching concern may be the desire of physicists to find a single idea that will bring together all of the particles and forces of nature into one grand unified theory. Steps have already been made in that direction by the unification of electrical and magnetic forces (the electromagnetic theory of James Clerk Maxwell in the late 1860s) and the unification of the electromagnetic and nuclear weak forces (the Electroweak Theory of Steven Weinberg, Abdus Salam, and Sheldon Glashow in the late 1960s). Physicists now feel that they are within reach of adding yet a third force—the strong nuclear force—to the Electroweak Theory, unifying three of the four basic forces in a Grand Unified Theory (GUT). But a unification of these three forces with the gravitational force still is well beyond their reach.

Albert Einstein spent a good share of the last years of his life looking for a single theory that would unify the four great forces of which he knew. He was unsuccessful, as have been all other physicists to date. However, for nearly two decades, physicists have been optimistic that an option is available that will not only unify the four basic forces, but also will resolve the questions of which particles are truly basic and how quantum mechanics and general relativity can be reconciled. That option is known as *superstring theory*, or, as it is more commonly known, *string theory*.

String Theory

The fundamental premise of string theory is that the basic units of matter are not point-like particles, such as quarks and leptons, but one-dimensional, line-like objects known as strings. Both the older notion of point-like particles and the newer notion of one-dimensional strings can be understood by comparison with their counterparts from Cartesian geometry. A point, for example, has no dimensions at all, although in physics it can have properties such as mass and spin. A string, by comparison, can be compared to a line, which has length but no thickness.

String theory consists of a complex collection of very complicated mathematical equations. It should not be expected that any beginning physics student would have even the vaguest understanding of the meaning of those equations or how they are manipulated mathematically. It is possible, however, to outline a few general conclusions that can be derived from the study of strings thus far.

In the first place, strings can be thought of, in a general way, as being somewhat similar to the strings on musical instruments such as the violin. They vibrate in space in much the way that violin strings vibrate when they are stroked or plucked. In fact, some of the mathematical analysis carried out in string theory uses the same mathematical equations developed for the analysis of vibrating strings more than a century ago.

Strings themselves are generally viewed as extraordinarily small objects with dimensions of no more than 10^{-33} cm. As a means of understanding an object this small, a string is about as large, compared with the proton in an atom, as the proton is compared with the size of the Solar System. Strings can also take more complex shapes than described in the preceding paragraph. For example, they may form loops that, themselves, may assume a variety of shapes. Each type of string has, of course, its own mode of vibration and, hence, unique characteristics.

One of the first conclusions to be derived from string theory is that much of what we already know about the structure of matter and the nature of energy can be described by the theory. For example, the particles that make up the Standard Model can all be thought of and represented as strings that vibrate with certain characteristic frequencies. A proton, for example, can be described as consisting of three strings, each representing a quark: two vibrating in one mode (the two up quarks in a proton) and the third vibrating in a second mode (the down quark in a proton).

During the first decade of its existence, string theory experienced some remarkable successes. It showed promise of being able to predict particles and conditions that appear to be necessary in devising a "theory of everything" that would include the four fundamental forces and all known particles. For example, the graviton, the force-carrying particle for gravitational energy, appeared as a necessary consequence in the solution of string equations.

Problems with String Theory

String theory is entirely a mathematical theory at this point. Those who work with the theory essentially manipulate mathematical equations to see what results they obtain. Sometimes those results are exciting and encouraging, as described above. At other times, they are frustrating and puzzling.

For example, one of the realizations that evolved in the early years of string theory was that strings required a space-time continuum of 26 dimensions. Most people have no trouble thinking of 3-dimensional space (in which we all live) or of 4-dimensional space in which time is factored with length, height, and depth. But space with 26 dimensions? Even sophisticated theoretical particle physicists had trouble with that one.

Over time, theorists found ways of dealing with this problem. They calculated that some spatial dimensions were "curled up" onto themselves into such small dimensions that it was not possible to observe them. In a way, this solution was a bit like "out of sight, out of mind." If one couldn't observe some of those dimensions, one didn't have to worry about their physical reality—as long as they continued to be taken into consideration in the relevant mathematical equations.

Today, physicists have reduced string theory to a condition in which 10 dimensions may be necessary, give or take a dimension. The "give or take" provision arises because modifications of string theory to solve new problems sometimes require the addition of another dimension. At this point, physicists appear willing to accept the necessity of the additional dimension in order to obtain a more important result, such as the integration of quantum mechanical and general relativistic principles. They'll figure out what the extra dimensions mean later on!

Another very difficult problem that developed for string theory was that at least five fundamentally different analytical approaches evolved. The five theories all started out with the assumption that strings exist and that they vibrate. But, beyond that point, they diverged in terms of the additional assumptions that became necessary and the conclusions that could be drawn. The question facing physicists became not, "Is string theory correct?" but, "Which of the five different theories, if any, is correct—*if* string theory in general is correct?"

M-Theory

The early 1990s saw the development of a promising modification to string theory known as M-theory. The "M" originally referred to *membrane*, but it has come to be taken by some physicists to represent *magical*, *mysterious*, *marvelous*—or, to detractors, *mythical*.

In M-theory, the fundamental particles of nature are neither dimensionless points nor 1-dimensional strings, but 2-dimensional membranes.

Imagine that you could slice a section out of a soap bubble and you would have a membrane of the kind treated in M-theory.

M-theory is certainly no simpler mathematically than string theory, nor does it resolve all the problems of string theory. For example, the existence of membranes immediately requires, in early calculations, the existence of one more dimension, and possibly two. Going from strings to membranes, then, means going from at least 10 to at least 11 or 12 dimensions. Not much progress there!

But M-theory does have the advantage of resolving the disagreement over five different string theories. It appears that the five versions of string theory can really be unified into a single theory by treating strings as forms of membranes rather than pure strings.

In fact, a string in M-theory is regarded as a 1-brane membrane. The term *brane* here refers to the number of dimensions involved, so that a 1-brane membrane is one with only one dimension, length. It is a string. Similarly, a 0-brane membrane is one with no physical dimensions, a point. Calculations show that membranes of many dimensions can exist. They are called, in general, p-brane membranes.

M-theory is in its very earliest stages of development, and as yet there is little understanding as to what it might and might not be able to do in integrating quantum mechanics and general relativity as well as the four forces of nature. But it has already produced some interesting and exciting results. For example, it may be able to explain the presence of Hawking radiation in black holes.

Black holes are objects in space that are formed when stars collapse upon themselves. They contain a very large concentration of mass in a very small volume. In fact, the gravitational force exerted by this mass is so great that nothing, including radiation energy (such as light), can escape. It is for that reason that they are called *black*.

Some years ago, however, the English astrophysicist Stephen Hawking demonstrated that very small amounts of radiation should still be able to escape from black holes. M-theory has been able to show how such radiation can be produced in, and then escape from, a black hole. The key to this approach has been to treat a black hole as a very large collection of 0-branes. When one analyzes the interaction among these 0-branes mathematically, the possibility of Hawking radiation appears as a part of the solution to the equations involved.

At this point, it is still too early to tell whether M-theory will be the pathway to solving some of the most important remaining fundamental issues in physics. For some, it is the magical road that may lead to that goal, while for others, it is still too mythical to take very seriously.

References

Garisto, Robert, "Curling Up Extra Dimensions in String Theory," <http://publish.aps.org/FOCUS/v1/st7.html> accessed 30 April 1999.

Kestenbaum, David, "Practical Tests for an 'Untestable' Theory of Everything?" *Science*, 7 August 1998: 758–59.

Mukhi, Sunil, "String Theory and the Unification of Forces," <http://theory.tifr.res.in/~mukhi/Physics/string.html> accessed 30 April 1999.

"The Official String Theory Web Site" <http://superstringtheory.com/index_p.html> accessed 30 April 1999.

Peterson, Ivars, "Loops of Gravity," *Science News*, 13 June 1998: 376–77.

Rensberger, Boyce, "A Short Course in String Theory (with No Equations)," *The Washington Post*, 11 December 1996; also at <http://www.washingtonpost.com/wp-srv/interact/longterm/horizon/121196/string.htm> accessed 30 April 1999.

Siegfried, Tom, "Physicists Sing Praises of Magical Mystery Theory," *The Dallas Morning News*, 28 October 1996: 8D.

"String Theory in a Nutshell" <http://www.strings.ph.qmw.ac.uk/WhatIs/Nutshell.html> accessed 30 April 1999.

"Superstrings? String Theory Home Page" <http://www.physics.ucsb.edu/~jpierre/strings/index.html> accessed 30 April 1999.

"Superstrings! String Theory Tutorial" <http://www.physics.ucsb.edu/~jpierre/strings/tutor.htm> accessed 30 April 1999.

Witten, Edward, "Duality, Spacetime and Quantum Mechanics," *Physics Today*, May 1997: 28–33.

FUSION POWER

Wait 'til next year! That phrase has long been used by supporters of athletic teams who fail to win a championship. But for more than 50 years, it has also been the rallying cry of supporters of nuclear fusion as a commercial source of energy.

Nuclear fusion is the process by which two or more small nuclei are forced together to produce one larger nucleus. The equation below is an extremely simplified representation of that process.

$$4 \ {}^1H \rightarrow {}^4He$$

This equation shows that four hydrogen nuclei are fused to produce a single helium nucleus.

The Promise of Fusion

The practical point of interest in these reactions is that a very small amount of mass is lost during the fusion process. That mass is converted to energy. It is not difficult to show mathematically that the energy released in a fusion reaction is, gram for gram, greater than that produced by any other known process. Fusion has, therefore, long been of great interest, both as a potential military weapon and as a source of energy for everyday applications such as the operation of industrial plants.

Fusion reactions were first hypothesized and studied more than seven decades ago as a mechanism by which stars produce energy. Shortly after the conclusion of World War II, scientists in both the United States and the then–Soviet Union confirmed that fusion reactions really are possible by building the first fusion (hydrogen) bombs. The incredible destruction caused by the first fusion bombs not only terrified the world, but also held out the promise for a bright new future powered by virtually limitless, inexpensive, fusion power. Optimistic nuclear physicists in the 1950s predicted that commercial fusion power would be available "within 10 years." Although that promise had still not been realized in the 1960s, nuclear physicists continued to reassert the "within 10 years" promise for another three decades.

The problem has been that finding a way to control a fusion reaction has been far more difficult than anyone ever imagined in the 1950s. Trapping the energy released by nuclear fission turned out to be "a piece of cake" compared with the similar challenge with nuclear fusion reactions. The first commercial power plant powered by atomic fission was the Yankee plant at Rowe, Massachusetts. It went on line in 1960 and was closed in 1991. The first commercial power plant powered by atomic fusion, by contrast, is still little more than a gleam in physicists' eyes. Indeed, after half a century of research, some authorities say that nuclear fusion will *never* be a practical method for generating power on a commercial level.

Technical Problems

The fundamental problem with which fusion experts must deal is electrostatics. The fusion of two hydrogen nuclei (protons) requires that two like-charged (positive) particles be brought into contact and then joined to each other. The energy required to bring about this change is very large, on the order of a few million degrees. Temperatures of this magnitude occur naturally in only one environment, the interior of stars. They can also be produced artificially in one other way, by the ignition of a nuclear fission weapon. The temperature at the center of such a weapon reaches a few million degrees for a fraction of a second after ignition. It is the explosion of a fission bomb, therefore, that provides the energy needed to initiate the fusion reactions that occur within a fusion bomb.

The first requirement in building a commercial fusion reactor, then, is to find a way of generating the very high temperature that is necessary to initiate fusion in the reactor fuel. That fuel usually consists not of ordinary hydrogen (protium), but of one or more heavier hydrogen isotopes, deuterium and/or tritium. Deuterium is an isotope of hydrogen whose nucleus contains one proton and one neutron, while tritium is an isotope whose nucleus contains one proton and two neutrons. Fusion reactions that involve deuterium and tritium occur more efficiently than those that involve protium only. All experimental fusion reactors today, then, use a mixture of deuterium and tritium as the fuel in which fusion occurs.

A second fundamental problem in designing a commercial fusion reactor is finding a way of containing the fusion reaction once it begins. No known material can withstand temperatures of a few million degrees, of course, so the question becomes one of finding a way to prevent the materials undergoing fusion from simply exploding away into the surrounding environment—in other words, to prevent them from turning into a fusion bomb.

The most common means for solving this problem has been magnetic confinement. Magnetic confinement is a method by which strong magnetic fields are wrapped around the materials used in a fusion reaction to prevent them from expanding beyond the fusion reactor. Magnetic confinement works because the fusion of hydrogen nuclei produces a hot form of ionized matter known as a *plasma*. A plasma is a mass of electrically charged particles whose movement can be manipulated by a magnetic field. The challenge for designers of fusion reactors has been to find exactly the right magnetic field shape to contain the plasma produced during a nuclear fusion reaction.

Fusion Reactor Design

Fusion research over the past 50 years has not produced a commercial reactor, but it has yielded an impressive amount of information about the nature of fusion reactions, the characteristics of plasmas, the technology of power generation, and the specific problems that must be solved to produce a break-even reactor. A *break-even* reactor is one that produces as much energy through fusion reactions as is needed to get those fusion reactions started. After five decades, researchers are fortunate to carry out an experiment in which even one-tenth of the break-even point is attained.

Nuclear scientists and engineers have created a number of designs for devices in which a sustainable fusion reaction might be produced. The most popular of these designs is called a tokamak. A *tokamak* is a doughnut-shaped ring surrounded by strong magnetic coils. Fusion is induced within the ring, and the plasma produced is held in place by the surrounding magnetic field. Tokamaks have been built at a variety of locations in the United States and around the world. Perhaps the most famous of the U.S. tokamaks has been the Tokamak Fusion Test Reactor (TFTR) in Princeton, New Jersey. Tokamaks have long been regarded as the most promising design for a successful commercial fusion reactor. The problem is that this promise has still not been achieved. In frustration, the U.S. Congress discontinued funding for TFTR in 1996, and that facility finally closed down in early 1997.

In light of the TFTR closure, U.S. physicists became even more committed to an international fusion project that had been on the drawing boards for nearly a decade, the International Thermonuclear Experimental Reactor (ITER). ITER was a joint project of the United States, Russia, the European Union (EU), and Japan, with primary funding coming from the EU and Japan. The project was expected to cost about $10 billion and was originally designed to begin construction in 1998 and to begin operation in 2008.

ITER was planned to be the largest tokamak ever built, with an oval-shaped ring 16 meters in diameter surrounded by a mass of magnets and electronic equipment 30 meters high. By late 1997, however, serious questions were being raised concerning the project. Economic crises in many parts of the world reduced confidence that scientists from participating nations could convince their governments that the project was worth the financial investment needed. The partner with the greatest investment, Japan, was going through a particularly difficult economic crisis, and even the United States, a relatively minor partner in the project, seemed unwilling to continue its modest financial contribution.

By late 1998, members of the ITER collaborative recognized that the original machine could not be built, and they began talking about a more modest project that would be half the size of the original reactor or that could be built for a few billion dollars less than the original estimate. Participants had given up the idea of trying to produce a break-even reactor and had decided to focus their efforts on learning more about the fusion process itself, including the properties of plasmas and the mechanisms by which energy can be generated and used to bring about fusion reactions.

Alternative Reactor Designs

Reactor designs other than the tokamak have also been suggested for fusion reactors. One of the earliest design concepts is called a stellarator. A *stellarator* consists of a number of spiral-shaped magnets wrapped around a central hollow ring in which the fuel is contained. Research on stellarators was largely ignored for many years. Then, in 1998, the largest machine of this design was put into operation at the National Institute of Fusion Science in Japan. Japanese scientists predict that it will function as

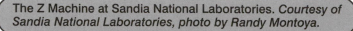

The Z Machine at Sandia National Laboratories. *Courtesy of Sandia National Laboratories, photo by Randy Montoya.*

efficiently as the defunct TFTR, although it will never produce enough energy to be used for commercial purposes.

Another new design being tested for fusion reactions is called the Z machine, located at the Sandia National Laboratories in Albuquerque, New Mexico. The Z machine makes use of an entirely new technology to produce and control the high temperature plasmas needed for fusion reactions to occur.

In this design, a parallel array of very thin wires is enclosed in a cylindrical container. A tiny pellet consisting of deuterium and tritium is placed at the center of the container. To initiate fusion, a massive electrical charge is passed through the wires, causing them to vaporize and implode (collapse inward). The vaporization and collapse of the wires generates a very large magnetic field, which compresses and shapes the plasma formed from the vaporized wires. At the same time, an enormous burst of X rays is produced by the plasma.

Sandia scientists have discovered that the proper arrangement of wires and input of energy can produce plasmas with the properties needed for fusion to occur. In a typical experiment, for example, 90 to 300 wires with diameters less than a human hair are enclosed in a cylindrical container 1.5 to 6.0 centimeters in diameter. The wires may be made from aluminum, tungsten, titanium, or some other metal. To generate the electrical power needed for plasma formation, 90,000 volts of power are used to charge a 5,000-cubic-meter capacitor, which then releases a 20-megampere (20 million ampere) current for a period of up to 100 nanoseconds.

The plasma produced in this reaction has a temperature of about 1.5 million degrees kelvin, roughly half that needed for a fusion reaction. The plasma also produces X rays with 40 terawatts (40 billion watts) of power, the most powerful X rays ever produced.

In an article on the Z machine, nuclear engineer Gerold Yonas reminisced about an earlier article he had written in 1978 about the promise of nuclear fusion, the expectation that a commercial fusion reactor would be available "within 10 years." He then admitted that the new Z machine encouraged him to believe that he could make that promise again—that the Z machine would make commercial fusion power available "within 10 years." At some point, the "wait 'til next year" promise for commercial fusion reactors may become a reality. Only time will tell.

References

Gibbs, W. Wayt, "A New Twist in Fusion," *Scientific American*, July 1998: 37.

Glanz, James, "Requiem for a Heavyweight at Meeting on Fusion Reactors," *Science*, 8 May 1998: 818–19.

Glanz, James, and Andrew Lawler, "Planning a Future without ITER," *Science*, 2 January 1998: 20–21.

Hellemans, Alexander, "JET Takes a Step Closer to Break-Even," *Science*, 3 October 1997: 29.

Lawler, Andrew, "Partners Will Rethink Fusion Project," *Science*, 30 January 1998: 649.

Peterson, Ivars, "The Z Machine," *Science News*, 17 January 1998: 46–47.

Yonas, Gerold, "Fusion and the Z Pinch," *Scientific American*, August 1998: 40–47.

CHAPTER FOUR
Biographical Sketches

I t is a persistent myth that a "great man" or "great woman" is behind the important breakthroughs in scientific research. Without a doubt, such individuals exist. Einstein, Feynman, Heisenberg, and Dirac are among the names that come to mind. The point that is easily missed, however, is that important steps in the advancement of physics and other sciences are often the result of many individuals working together.

No biographical chapter in a book on important advances in physics can, therefore, be completely fair to all those whose work has contributed to such advances. The biographies that have been chosen for this chapter meet one of two criteria. First, they are the stories of giants in the field who have been recognized for their accomplishments by the Nobel Prize. Second, they summarize the life of leaders or important participants in research teams whose work is summarized in Chapters 1 and 2.

Satyendra Nath Bose (1894–1974)

Satyendra Bose is perhaps best known for his analysis of Planck's Law of Black Body Radiation using Einstein's concept of the atomic nature of electromagnetic radiation. This analysis led to the concept of a mass of particles (such as atoms) all having the same quantum state. Predicted by Bose in a 1924 paper, the existence of such a state was finally confirmed experimentally in 1995 by physicists at the JILA in Boulder, Colorado.

Bose was born in Calcutta, India, on 1 January 1894. He began his education in English-language schools in Calcutta, but then transferred to Bengali-language schools during the rise of Indian nationalism in the early 1900s. He studied science at the Presidency College, from which he received B.Sc. and M.Sc. degrees in 1913 and 1915.

Bose was appointed a lecturer at the University College in Calcutta in 1916 and was then appointed a reader in physics at the University of Dacca in 1921. He later traveled to Paris to work with Marie Curie, and to Berlin where he studied with Albert Einstein. Bose returned to Calcutta in 1926 to become Khaira Professor of Physics, a post he held until 1958. He was elected a Fellow of the Royal Society in 1958. Bose died on 4 February 1974 in Calcutta.

Jean-Claude Bradley (1968–)

Jean-Claude Bradley's primary field of research involves the invention of methods for producing contactless three-dimensional electrical circuits. This work is vital in the development of techniques for manufacturing nanoscale electronic devices.

Bradley was born in Joliette, Quebec, Canada, on 17 November 1968. He received a B.S. in chemistry in 1989 from Laurentian University in Sudbury, Ontario, and a Ph.D. in 1993 in synthetic and mechanistic organic chemistry from the University of Ottawa. He says that he was then "drawn to warmer climates" and carried out his postdoctoral studies on photolithographic synthesis of DNA arrays at Duke University in 1993–94. A second postdoctoral program was completed over the next two years at the Collège de France, in Paris.

"Missing the sound of the English language," Bradley then returned to the United States where he accepted a position as assistant professor of chemistry at Drexel University in Philadelphia.

Bradley has received a number of honors and awards including the Elizabeth Burton Award for 1986–87, the CIC Silver Medal in Chemistry in 1988, and the Silver Medal of the Governor-General of Canada in 1989, all at Laurentian University; University of Ottawa and NSERC scholarships in 1989–92 and 1989–93, respectively, at the University of Ottawa and at the Bourse de l'Association Française pour la Lutte at the Collège de France in 1994–95.

Steven Chu (1948–)

Steven Chu was selected to receive a share of the 1997 Nobel Prize in Physics for his work on methods for using laser light to keep gases contained in "traps" cooled to temperatures close to absolute zero

(–273.15°C). These methods have provided new and unique ways to study the quantum properties of large numbers of atoms in a common quantum state.

Chu was born in St. Louis, Missouri, on 28 February 1948. He was awarded a B.S. in physics and an A.B. in mathematics by the University of Rochester in 1970 and a Ph.D. in physics by the University of California at Berkeley in 1976.

Chu remained at Berkeley as a postdoctoral fellow for two years, then served as a member of the technical staff at the Bell Laboratories in Murray Hill, New Jersey, from 1978 to 1983. In 1983, he was appointed head of the Quantum Electronics Research Department at the AT&T Bell Laboratories in Holmdel, New Jersey. Chu left AT&T in 1987 to become professor of physics and applied physics at Stanford University, a post he currently holds.

Chu has been honored with the Broida Prize for Laser Spectroscopy in 1987, the APS/AAPT Richtmyer Memorial Prize in 1990, the King Faisal International Prize in Science (co-winner) in 1993, the American Physical Society Arthur L. Schawlow Prize in Laser Science in 1994, and the Optical Society of America's William F. Meggers Award for Spectroscopy in 1994. He has also been made a Fellow of the American Academy of Arts and Sciences and a member of the National Academy of Sciences and the Academia Sinica.

Claude Cohen-Tannoudji (1933–)

Claude Cohen-Tannoudji shared the 1997 Nobel Prize in Physics with Steven Chu and William D. Phillips for their research on methods for cooling and trapping atoms by using laser light. This technique has made it possible to immobilize and study relatively large groups of atoms with identical quantum properties. As a result of this research, physicists will have an insight on the nature of matter that has never before been available.

Cohen-Tannoudji was born in Constantine, Algeria, in 1933. He received a Ph.D. in physics from the Ecole Normale Supérieure at Paris in 1962. In 1973, he was appointed professor at the Collège de France, in Paris. He is currently affiliated with the Laboratorie de Physique at the Ecole Normale Supérieure. He was elected to membership in the French Academy of Science in 1981 and has been awarded many honors for his work on laser cooling and trapping of atoms. Among these are the 1996 Quantum Electronics Prize of the European Physical Society.

Eric A. Cornell (1961–)

Eric Cornell's field of special interest is laser cooling, including the production and study of Bose-Einstein condensates. He and his colleagues at the Joint Institute for Laboratory Astrophysics (JILA) in Boulder, Colorado, were the first researchers to report on the production of such a state.

Cornell was born in Palo Alto, California, on 19 December 1961. He was awarded a B.S. in physics, with honors and with distinction, by Stanford University in 1985 and a Ph.D. in physics by the Massachusetts Institute of Technology in 1990.

Cornell's teaching experience began with a post as teacher of English as a Foreign Language at the Taichung YMCA in Taiwan in 1982. He was a research assistant at Stanford from 1982 to 1985 and a teaching fellow at the Harvard Extension School in 1989. After earning his doctorate, Cornell spent two years as a postdoctoral student at JILA, and then was appointed assistant adjoint professor (1992–95) and adjoint professor (1995–present) at the University of Colorado (UC) at Boulder. He has also been a staff scientist at the National Institute of Standards and Technology (NIST) in Boulder from 1992 to the present and a Fellow at JILA, NIST, and UC from 1994 to the present.

Among his honors and awards are the Samuel Wesley Stratton Award in 1995; the Newcomb-Cleveland Prize in 1995–96; the Carl Zeiss Award in 1996; the Fritz London Prize in Low Temperature Physics in 1996; the I. I. Rabi Prize in Atomic, Molecular and Optical Physics in 1997; the King Faisal International Prize in Science in 1997; and the National Science Foundation's Alan T. Waterman Award in 1997. He is the author and coauthor of more than 50 articles and conference reports.

Robert F. Curl Jr. (1933–)

Robert F. Curl Jr. received the Nobel Prize for Chemistry in 1996, along with Richard Smalley and Harold Kroto, for his discovery of the form of carbon known as buckminsterfullerene, or, more simply, as buckyballs or fullerenes. This allotrope of carbon consists of soccer- or rugby-ball-shaped molecules of 60 or more carbon atoms. The discovery of the fullerenes set into motion a dramatically new direction for chemical research that has led to amazing nanostructures with apparently unlimited applications in material sciences, electronics, and other fields.

Curl was born in Alice, Texas, on 23 August 1933. He claims that his interest in chemistry can be traced to having received a chemistry set as a Christmas present at the age of nine. He chose to attend Rice Institute for his undergraduate work in chemistry, apparently because of the institution's fame as a football power. He earned his B.A. in chemistry at Rice in 1954 and then was awarded a Ph.D. in chemistry at the University of California at Berkeley in 1957.

After a year at Harvard, Curl returned to Rice, where he was appointed assistant professor of chemistry. Except for sabbatical leaves, he has remained at Rice ever since, becoming a full professor in 1967 and serving as chair of the department of chemistry from 1992 to 1996.

In addition to his Nobel Prize, Curl has been awarded the Clayton Prize by the Institute of Mechanical Engineers in 1957 and, with Kroto and Smalley, the American Physical Society International Prize of New Materials in 1992.

Cees Dekker (1959–)

Cees Dekker's special fields of interest include molecular electronics, scanning-probe microscopy, and nanotechnology. In 1997, his research team reported on the development of a resistor consisting of a single nanotube. Such devices will play critical roles in the future development of technology at the molecular level.

Dekker was born in Haren, the Netherlands, on 7 April 1959. He earned a master's degree in experimental physics and a Ph.D. in physics from the University of Utrecht in 1984 and 1988, respectively. During the period from 1984 to 1988 he worked as research assistant in the solid state physics group of H. W. de Wijn at Utrecht. After receiving his doctorate, he was appointed to the permanent scientific staff in the solid state physics group of R. J. J. Zijlstra. Since 1993, he has worked in the permanent scientific staff of the quantum transport group of J. E. Mooij at Delft University. During the academic year 1990–91, Dekker was visiting researcher at the IBM Research Center in Yorktown Heights, New York.

Amilra Prasanna de Silva (1952–)

Amilra Prasanna de Silva's area of special interest involves the design of large molecules with the capability of receiving and sending signals by means of fluorescent and luminescent mechanisms. In 1997, a research team under his direction reported on the development of the first molecular AND logic gate.

De Silva was born in Colombo, Sri Lanka, on 29 April 1952. He attended the University of Colombo, from which a received a B.Sc. with First Class Honours in 1975. He was assistant lecturer in chemistry at the University of Sri Lanka, Colombo campus, for one year (1975–76) before continuing his education at Queen's University in Belfast, Ireland. In 1980, he was awarded a Ph.D. in organic photochemistry from Queen's.

In 1980, de Silva returned to Sri Lanka and took a post as lecturer in chemistry at the University of Colombo. After six years in that post, he returned to Queen's as lecturer in chemistry, a position he held until his appointment as reader in chemistry in 1991. In 1997, de Silva was also appointed to his current post as professor of chemistry and reader in chemistry at Queen's University.

Daniel Herschel Eli Dubin (1956–)

Daniel Dubin and one of his doctoral students, Dehze Jin, have studied the formation of stable geometric structures within apparently chaotic systems. Their work has provided a new insight on a fundamental aspect of nature, entropy.

Dubin was born in Cochrane, Ontario, Canada, on 13 September 1956. He earned a B.Sc. degree, with honors, in theoretical physics, at Queen's University in Kingston, Ontario, in 1978. He was then awarded a Ph.D. in astrophysics by Princeton University in 1984.

Dubin's academic experience includes posts as experimental research assistant and theoretical research assistant at the Princeton Plasma Physics Laboratory (1979–84) and postgraduate researcher (1984–86), assistant research physicist (1986–87), assistant professor (1987–94), and associate professor (1994–present) at the University of California at San Diego.

Among the honors and awards Dubin has received are the Susan Near Prize at Queen's University (1977), the Maj. J. H. Rattray Scholarship at Queen's (1977), and the Natural Sciences and Engineering Research Council of Canada Postgraduate Scholarship at Princeton (1981–83). He was elected a Fellow of the American Physical Society, Division of Plasma Physics, in 1994. He is the author or coauthor of more than a dozen peer-reviewed papers.

Albert Einstein (1879–1955)

Albert Einstein is generally regarded as one of the greatest physicists who ever lived. Among his many contributions to the field of science was his analysis of the possibility of there being large groups of particles (such as atoms), all with the same quantum characteristics. The theoretical basis for the existence of such groups—now known as Bose-Einstein condensates—was developed by Einstein and the Indian physicist Satyendra Bose in the 1920s.

Einstein was born in Ulm, Germany, on 14 March 1879. He began his secondary school in Milan, where he did so poorly that he was asked to leave school. After much difficulty, Einstein was admitted to the Eidgenössiche Technische Hochschule (ETH; now the Swiss Federal

Institute of Technology), where he received his bachelor's degree in teaching in 1900. Because of his poor school record, he was unable to find a teaching job and, instead, took a post at the Swiss Patent Office in Bern. He remained in this job for seven years, the most productive period of his professional life.

Three of Einstein's most important scientific contributions were all made in a single year, 1905. During this period, he published papers on Brownian movement, on the photoelectric effect, and on the Special Theory of Relativity.

Despite these contributions, fame came slowly to Einstein. He finally received his doctoral degree from the University of Zurich in 1905, after which he spent a year teaching at the University of Bern. He later held academic appointments at the University of Zurich, the German University of Prague, ETH, and the Kaiser Wilhelm Institute for Physics in Berlin.

By 1933, Einstein had become convinced of the personal dangers he faced with the rise of Nazism in Germany. He accepted an appointment at the Institute for Advanced Studies in Princeton, New Jersey, a post he held until his death on 18 April 1955 in Princeton.

Stephen Elliott (1952–)

Stephen Elliott was the leader of a research group at the University of Cambridge who reported in 1997 on the preparation of an anisotropic type of glass. This development holds important potential applications in the development of nanoscale electronic devices.

Elliott was born on 30 September 1952 in England. He earned a B.A. degree from Trinity College, Cambridge, in 1974, and a Ph.D. degree at the Cavendish Laboratory and Trinity College in Cambridge in 1977. He was demonstrator in the department of physical chemistry at Cambridge from 1979 to 1984, and was then promoted to lecturer in physical chemistry in 1984 and then to reader in solid-state chemical physics at Cambridge in 1994. In 1998, he also accepted an appointment as professor in the department of physics at the Ecole Polytechnique in Palaiseau, France.

In 1992 Elliott was awarded the Zachariasen Prize for the researcher under the age of 40 who has made the most significant and innovative advance in the field of noncrystalline materials. He serves on a number of committees, including the International Advisory Board of the International Conference on Solid-State Chemistry, the International Advisory Board of the International Conference on Non-Crystalline Materials, and the International Advisory Board of the International Conference on Amorphous Semiconductors.

Murray Gell-Mann (1929–)

Murray Gell-Mann was awarded the 1969 Nobel Prize in Physics for his development of a system for classifying subatomic particles. By the early 1960s, the number of so-called fundamental particles was very large, in excess of 200, much too large for most physicists to consider them all truly fundamental. Gell-Mann developed a scheme for classifying these particles on the basis of a small number of truly basic particles, now known as quarks and leptons. His work forms the basis of the Standard Model by which elementary particles are still classified.

Gell-Mann was born in New York City on 15 September 1929. He was a brilliant student and entered Yale University at the age of 15. He graduated with a B.S. degree in physics in 1948 and then entered the Massachusetts Institute of Technology (MIT). He was awarded a Ph.D. by MIT in 1951.

Gell-Mann's first academic appointment was at the Institute for Advanced Studies in Princeton, New Jersey. After two years there, he accepted an appointment as associate professor of physics at the University of Chicago. In 1955, he left Chicago to become associate professor of physics at the California Institute of Technology (Caltech). In 1966, he was appointed R. A. Millikan Professor of Theoretical Physics at Caltech.

In addition to the Nobel Prize, Gell-Mann has been awarded the Dannie Heineman Prize of the American Physical Society, the E. O. Lawrence Memorial Award for Physics, the Franklin Medal of the Franklin Institute, and the National Academy of Science's John J. Carty Medal.

Alan Guth (1947–)

Alan Guth is often referred to as the Father of Inflationary Theory. Inflationary theory is an explanation of the changes that took place in the universe in the first few seconds after its birth. This explanation is of enormous importance because of the implications it holds for the properties and behavior of the universe today. Current research has recently begun to produce evidence against which the inflationary theory can be tested.

Guth was born on 27 February 1947 in New Brunswick, New Jersey. He received B.S. and M.S. degrees in physics from the Massachusetts Institute of Technology (MIT) in 1969 and then remained at MIT for his doctorate, which was awarded in 1972.

Guth's first academic appointment was at Princeton University, where he was instructor of physics from 1971 to 1974. He was then appointed a postdoctoral research associate at Columbia University from 1974 to 1977. From 1977 to 1979, he held a similar appointment at Cornell

University. After a year's sabbatical at the Stanford Linear Accelerator Center at Stanford University, Guth received an appointment at MIT. He was promoted through the ranks at MIT until he reached the position of Jerrold Zacharias Professor of Physics, a post he continues to hold. Between 1984 and 1990, Guth was also on the staff of the Harvard-Smithsonian Center for Astrophysics.

Peter Ware Higgs (1929–)

Peter Higgs has made important contributions to the field of elementary particle physics. His name is memorialized in a hypothesized subatomic particle, the Higgs boson, that physicists are hoping to discover when more powerful particle accelerators become available. Among other properties, the Higgs boson is believed to be responsible for the property of mass that all matter demonstrates.

Higgs was born in Newcastle upon Tyne on 29 May 1929. He attended King's College at the University of London, where he received his B.Sc., M.Sc., and Ph.D. degrees in physics in 1950, 1951, and 1954, respectively. After earning his doctorate, Higgs was appointed senior research fellow at the University of Edinburgh for one year (1955–56). He was then I.C.I. Research Fellow at the University of London from 1956 to 1958. From 1959 to 1960, he was temporary lecturer in mathematics at University College, London. In 1960, Higgs was appointed lecturer in mathematical physics at the University of Edinburgh, and then became reader in mathematical physics (1970–80), professor of theoretical physics (1980–96), and finally, professor emeritus.

Among Higgs's many awards are the Hughes Medal of the Royal Society (1981), the Rutherford Medal of the Institute of Physics (1984), the Scottish Science Award of the Saltire Society and Royal Bank of Scotland (1990), the Paul Dirac Medal and Prize of the Institute of Physics (1997), and the High Energy and Particle Physics Prize of the European Physical Society (1997).

Steven J. Hillenius (1951–)

Steven Hillenius was lead researcher on the research team at Bell-Lucent Laboratories that reported in 1998 on the development of the smallest transistor ever made, one with an insulating layer only three atoms thick. This research was important not only because of the breakthrough in size it represents, but also because of the new insights it provides on problems that will have to be solved in making nanosize electronic components.

Hillenius was born in Hackensack, New Jersey, on 10 October 1951. He was awarded a B.S. in physics by the University of Delaware in 1973 and a Ph.D. in physics by the University of Virginia in 1979. He then accepted an appointment as professor of physics at the University of Virginia, where he served from 1978 to 1981. His area of special interest there was fundamental research in low temperature solid state physics.

In 1981, Hillenius left Virginia to join Bell Laboratories, Lucent Technologies, in Murray Hill, New Jersey, as a member of the technical staff. In 1984, he was assigned to the Bell Laboratories' VLSI Technology Development Group, and in 1989, he was made manager of Bell's CMOS Technology Development Group. In 1991, he was appointed Laboratories Research Technical Manager of the Device Research Group, a post he held until 1996. He was then made head of the ULSI Technology Research Department, a position he continues to hold.

In 1996, Hillenius was elected to the grade of IEEE (Institute of Electrical and Electronics Engineers) Fellow in recognition of his contributions to the field of solid-state technology and its applications to integrated circuits. His outside interests include whitewater boating, skiing, and hiking.

Edwin Powell Hubble (1889–1953)

Edwin Hubble is widely regarded as one of the most important observational astronomers in history. Hubble's fame rests on two major discoveries. The first of those discoveries was that the universe extends far beyond our own galaxy and includes untold numbers of galaxies like our own. The second discovery was that the universe is not static but, instead, is constantly expanding. His calculation of the rate of this expansion is now known as Hubble's Law.

Hubble was born in Marshfield, Missouri, on 20 November 1889. He attended the University of Chicago, from which he received a B.S. in mathematics and astronomy in 1910. He was then awarded a Rhodes Scholarship and attended The Queen's College at the University of Oxford from 1910 to 1912. He studied law at Oxford and returned to the United States in 1913 to open a law office in Louisville, Kentucky.

Hubble soon became bored with law and decided to return to the University of Chicago to pursue a doctoral program in astronomy. He received a Ph.D. in that field in 1916 and was invited to join the staff at the Mount Wilson Observatory near Pasadena, California. After a short delay because of World War I, Hubble accepted that appointment. He remained at the Mount Wilson Observatory for the rest of his professional career except for another interruption during World War II. Hubble died of a cerebral thrombosis in Pasadena on 28 September 1953.

Sumio Iijima (1939–)

Sumio Iijima was the first researcher to observe the formation of carbon nanotubes in the soot produced during the electric-arc vaporization of carbon. This discovery occurred in 1991 when Iijima was employed at the NEC Corporation in Japan.

Iijima was born in Japan in 1939 and received his undergraduate education at the University of Electro-Communications in Tokyo. He then transferred to Tohoku University, in Sendai, Japan, where he earned a Ph.D. in physics. He worked for 12 years, from 1970 to 1982, in the field of high resolution electron microscopy at Arizona State University, first as a postdoctoral student, and later as a research scientist. Iijima also spent a year at Cambridge University in 1979, working on the structure of graphite as revealed through electron microscopy.

In 1982, Iijima returned to Japan to join the ERATO Ultrafine Particles Project. Five years later, he moved to NEC as a research fellow. Among Iijima's many honors are the 1996 Asahi Award, the Hishina Memorial Award, and the B. E. Warren Award.

Dezhe Jin (1967–)

Dezhe Jin has received recognition for his study of the formation of stable geometric patterns in electron plasmas. This work has provided an interesting insight on the development of ordered structures in systems that appear to be chaotic and disordered.

Jin was born in Tumen City, a small town in northeast China, in February 1967. He earned B.S. and M.S. degrees in physics at Tsinghua University in Beijing in 1990 and 1994, respectively. He then entered a doctoral program at the University of California at San Diego (UCSD), where he was awarded a Ph.D. in physics in 1999. Jin's teaching experience has included positions as assistant lecturer at Tsinghua University and teaching assistant at UCSD.

Jin's leisure activities include singing, songwriting, and guitar playing. He has recorded and produced one album of his own music, "Those Were the Dreams."

John D. Joannopoulos (1947–)

John Joannopoulos was leader of a research team at the Massachusetts Institute of Technology (MIT) that announced the invention of a photonic crystal in 1997. This invention represents an important breakthrough in the technology for using light waves rather than electrons for the transmission of messages.

Joannopoulos was born on 26 April 1947 in New York City. He received B.A. and Ph.D. degrees from the University of California at Berkeley in 1968 and 1974, respectively. He was then offered a position as assistant professor of physics at MIT in 1974. He was later promoted to associate professor in 1978 and to professor in 1983, a post he continues to hold. Joannopoulos has been an Alfred P. Sloan Fellow, a John S. Guggenheim Fellow, and an American Physical Society Fellow, in addition to earning a G.S.C. School of Science Graduate Testing Award in 1991.

Joannopoulos serves on a number of professional committees and boards, including the International Conference on Physics and Chemistry of Solids (1978–81), the International Conference on Amorphous and Liquid Semiconductors (1979–80), and the International Conference on the Physics of Semiconductors (1982–84 and 1988–90).

Henry Cornelius Kapteyn (1963–)

Henry Kapteyn was leader of the research team at the University of Michigan that announced the invention of an X-ray laser in 1998. The invention is especially important because it significantly expands the number of circumstances in which laser technology can be utilized.

Kapteyn was born in Oak Lawn, Illinois, on 21 January 1963. He earned a B.S. degree in physics from Harvey Mudd College in 1982, an M.A. in physics from Princeton University in 1984, and a Ph.D. in physics from the University of California at Berkeley in 1989.

After graduation, Kapteyn remained at Berkeley for a one-year postdoctoral program. He then accepted an appointment as assistant professor of physics at Washington State University in 1990. He was promoted to associate professor there in 1995 before accepting an appointment as associate professor of electrical engineering at the University of Michigan in 1996, a post he still holds.

Kapteyn has been awarded the Adolph Lomb Medal of the Optical Society of America (1993) and was elected a Fellow of the Optical Society of America in 1998. He is the author or coauthor of more than 100 papers and book chapters. He is also the cofounder and president of Kapteyn-Murnane Laboratories, manufacturers of ultrafast laser instruments.

Wolfgang Ketterle (1957–)

Wolfgang Ketterle and his colleagues at the Massachusetts Institute of Technology (MIT) have conducted seminal research on the production and study of Bose-Einstein condensates and were the first researchers to report on the development of an atomic laser in 1995.

Ketterle received a diploma (equivalent to a master's degree) from the Technical University of Munich in 1982 and a Ph.D. in physics from the University of Munich in 1996. He then completed postdoctoral work at the Max-Planck Institute for Quantum Optics in Garching, at the University of Heidelberg, and at MIT. He became a member of the faculty at MIT in 1993 and is now John D. MacArthur Professor there.

Among Ketterle's honors and awards are a David and Lucille Packard Fellowship in 1996, the Rabi Prize of the American Physical Society in 1997, the Gustav-Hertz Prize of the German Physical Society in 1997, and the Discover Magazine Award for Technological Innovation in 1998. He was also selected as Distinguished Traveling Lecturer of the Division of Laser Science of the American Physical Society in 1998–99.

Daniel Kleppner (1932–)

Daniel Kleppner and his colleague, Thomas Greytak, announced the discovery in 1998 of a Bose-Einstein condensate of hydrogen. This discovery marked the culmination of two decades of research on this fascinating new form of matter in which all particles exist in the same quantum state.

Kleppner was born on 16 December 1932 in New York City. He received a B.S. from Williams College in 1953, a B.A. from Cambridge University in 1955, and a Ph.D. from Harvard University in 1959. Kleppner's first academic appointment was as assistant professor of physics at Harvard from 1962 to 1966. He then moved to the Massachusetts Institute of Technology (MIT), where he became associate professor of physics (1966–73), professor of physics (1974–present), and Lester Wolfe Professor of Physics (1985–present). He has also served as associate director of the Research Laboratory of Electronics at MIT since 1987.

Kleppner was awarded the Davisson-Germer Prize of the American Physical Society in 1985, the Julius Edgar Lilienfeld Prize of the American Physical Society in 1990, and the William F. Meggers Award of the Optical Society of America in 1991. He has been guest lecturer at Harvard University, Rice University, Pennsylvania State University, Oregon State University, and Hamline University.

Harold Kroto (1939–)

Harold Kroto was born in Wisbech, Cambridgeshire, England. He attended the University of Sheffield, from which he received a bachelor's degree in chemistry in 1961 and a Ph.D. in chemistry in 1964. After two years of postdoctoral research at the National Research Council in Ottawa, Canada, and one year at Bell Laboratories in New Jersey, Kroto accepted an appointment at the University of Sussex in Brighton in 1967.

He became full professor at Sussex in 1985, and in 1991, he was made Royal Society Research Professor at the university.

In 1985, Kroto, Richard E. Smalley, and Robert F. Curl Jr. combined in the discovery of a new allotropic form of carbon known as buckminsterfullerene, a soccer-ball-shaped sphere consisting of 60 carbon atoms joined to each other in pentagons and hexagons. For this work, Kroto, Smalley, and Curl were awarded the 1996 Nobel Prize for Chemistry. In the same year, Kroto was knighted by Queen Elizabeth II.

In addition to his Nobel Prize, Kroto has been named Tildon Lecturer of the Royal Society of Chemistry (1981–82); was elected a Fellow of the Royal Society (1990); and has been awarded the International Prize for New Materials of the American Physical Society with Smalley and Curl (1992), the Italgas Prize for Innovation in Chemistry by the Université Libre de Bruxelles (1992), the Longstaff Medal of the Royal Society of Chemistry (1993), the Hewlett Packard Europhysics Prize (1994), and the Moet Hennessy/Louis Vuitton Science pour l'Art Prize (1994). He has also received honorary doctoral degrees from the University of Sheffield and the University of Kingston, both in 1995.

Robert B. Laughlin (1950–)

Robert Laughlin was awarded a share of the 1998 Nobel Prize in Physics for his contribution to the understanding of the fractional quantum Hall effect. This effect was first discovered by co-winners Horst L. Störmer and Daniel C. Tsui in 1982. It provided dramatic new evidence about the properties of a quantum fluid with potentially important applications in the field of semiconductor fabrication.

Laughlin was born in 1950 in Visalia, California. He earned an A.B. degree in 1972 from the University of California at Berkeley and a Ph.D. in 1979 from the Massachusetts Institute of Technology. He was appointed in 1982 to the position of research physicist at the Lawrence Livermore National Laboratory, a post he continues to hold. In 1985, he was also appointed associate professor of physics at Stanford University. He was promoted to professor of physics in 1989, a post he also continues to hold.

In addition to the Nobel Prize, Laughlin has received the E. O. Lawrence Award for Physics (1985) and the Oliver E. Buckley Prize (1986). He has been elected a Fellow of the American Academy of Arts and Sciences, of the American Association for the Advancement of Science, and of the National Academy of Sciences.

David M. Lee (1931–)

David Lee was awarded a share of the 1996 Nobel Prize in Physics for his work on superfluid helium-3. Helium-3 is an isotope of helium that takes on dramatically unusual properties at temperatures approaching absolute zero (−273.15°C). These properties become manifest as the isotope loses all inner friction and begins to flow upward, around, and through spaces where it would normally not travel. This behavior cannot be explained by classical physics, only by the application of quantum mechanical principles.

Lee was born in Rye, New York, on 20 January 1931. He received a B.A. degree, cum laude, from Harvard University in 1952, an M.S. from the University of Connecticut in 1955, and a Ph.D. from Yale in 1959. He served in the U.S. Army from 1952 to 1954.

Lee has spent all of his academic career at Cornell University, where he has served successively as instructor, assistant professor, associate professor, and professor. He has also been visiting scientist or visiting professor at the Brookhaven National Laboratory, the University of Florida, Peking University (China), and the University of California at San Diego. In addition to his Nobel Prize, Lee has been awarded the Sir Francis Simon Memorial Prize (1976) and the Oliver E. Buckley Condensed Matter Physics Prize of the American Physical Society (1981).

Andrei Linde (1948–)

Andrei Linde is one of the authors of the inflationary universe scenario. That scenario deals with changes that took place in the first few seconds after the birth of the universe. He has also contributed to an understanding of the Higgs boson, suggesting an upper limit for the mass of that hypothesized particle in 1976.

Linde was born in Moscow, Russia, on 2 March 1948. He studied physics at Moscow State University from 1966 to 1971 and then became a graduate student at the Lebedev Physical Institute in Moscow. He earned a doctoral degree in 1974 on cosmological phase transitions.

Linde's contribution to inflationary theory involved corrections in the amount of time involved in the early stages of creation. Working with Paul Steinhardt and Andreas Albrecht in 1984, Linde was able to devise equations that resolved this problem.

Linde is currently professor of physics at Stanford University. He is the author of 150 papers and two books on particles physics and cosmology. In 1978, he was awarded the Lomonosov Prize of the Academy of Sciences of the former Soviet Union.

Robert M. Metzger (1940–)

Robert Metzger's special field of interest is the conductivity of single organic crystals and the possibility of using such crystals as molecular rectifiers. He reported on the apparent development of such a device in 1997.

Metzger was born in Yokohama, Japan, on 7 May 1940. He received a B.A. in 1962 from the University of California at Los Angeles and a Ph.D. from the California Institute of Technology in 1968. His postdoctoral work was done at Stanford University from 1969 to 1971.

Metzger has been professor of chemistry at the University of Alabama since 1986. He came to Alabama from the University of Mississippi, where he had been assistant professor (1971–76), associate professor (1976–82), professor (1982–84), and Margaret McLane Coulter Professor (1984–86). He has also been Gast Professor at the University of Heidelberg (1979–80) and Maitre de Conferences Associe at the University of Bordeaux (1980).

Walter Munk (1917–)

In 1979, Walter Munk, along with Carl Wunsch, first suggested the idea of using sound waves to study the properties of the oceans. That idea has been realized in the Acoustic Thermometry of Ocean Climate (ATOC) project.

Munk was born in Vienna, Austria, on 19 October 1917, and moved to New York City at the age of 14. He studied physics at Columbia University and at the California Institute of Technology, from which he earned B.S. and M.S. degrees. He received his Ph.D. in oceanography from Scripps Institution of Oceanography in 1947. During World War II, Munk served in the U.S. Army Ski Battalion. He was also an oceanographer with the University of California Division of War Research and a meteorologist for the Army Air Force. In 1947, Munk joined the faculty of the University of California at San Diego (UCSD) as assistant professor. He later was promoted to associate professor and, in 1954, became professor of geophysics. In 1955, he founded the University of California Institute of Geophysics and Planetary Physics and served as director of that institute until 1984. He is currently professor emeritus of oceanography at UCSD.

Taku Okuchi (1969–)

While still a graduate student, Taku Okuchi offered a revolutionary new explanation regarding the composition of the Earth's core. He suggested that hydrogen constitutes a significant portion of the core, a concept that had rarely been seriously considered previously, and one that could radically change scientists' views of the structure and formation of the Earth.

Okuchi was born in Tokyo on 15 May 1969. He was awarded a bachelor's degree in 1993 from the Department of Geology and Mineralogy of Kyoto University. He then enrolled at the Tokyo Institute of Technology, from which he received a Ph.D. in 1998. He then accepted appointments as a research fellow at the Institute of Low Temperature Science at Hokkaido University and as a research assistant in the Department of Earth and Planetary Sciences at Nagoya University, positions he continues to hold.

Okuchi is a member of the Japanese Society for Planetary Science, the Japan Society of High Pressure Science and Technology, and the American Geophysical Union.

Douglas D. Osherhoff (1945–)

Douglas Osherhoff earned a share of the 1996 Nobel Prize in Physics for his research on superfluid helium-3. Superfluidity is the tendency of certain liquids to flow in ways unlike those of typical liquids. For example, a superfluid liquid flows up the sides of a container and through very small holes that would normally impede its passage. Osherhoff and his co-winners, David Lee and Robert Richardson, discovered the superfluid behavior of helium-3 unexpectedly in the early 1970s.

Osherhoff, a graduate student at the time of the discovery, was born in Aberdeen, Washington, on 1 August 1945. He earned a B.S. from the California Institute of Technology in 1967 and a Ph.D. from Cornell in 1973. He served as a member of the technical staff at the AT&T Bell Laboratories from 1972 to 1987, during which time he held the post of head of the Solid State and Low Temperature Research Department from 1981 to 1987. In the latter year, he was appointed professor of physics and applied physics at Stanford University, where he is now J. G. Jackson and C. J. Wood Professor of Physics.

In addition to the Nobel Prize, Osherhoff has been awarded the Simon Memorial Prize (1976) and the Oliver E. Buckley Prize (1981). He has been named a MacArthur Prize Fellow (1981), and a Fellow of the National Academy of Sciences, the American Physical Society, and the American Academy of Arts and Sciences.

William D. Phillips (1948–)

William Phillips was awarded a share of the 1997 Nobel Prize in Physics for his research on the development of techniques for immobilizing and studying groups of atoms at very low temperatures. These techniques make use of temperatures very close to absolute zero (–273.15°C) and laser beams to "trap" atoms in a small volume, making it possible to study their properties in detail. These properties are very different from those observed at higher temperatures and can be explained only by the application of quantum mechanical principles.

Phillips was born on 5 November 1948 in Wilkes-Barre, Pennsylvania. He attended Juniata College, from which he received a B.S. in physics in 1970. He then continued his studies at the Massachusetts Institute of Technology (MIT), where he earned a Ph.D. in physics in 1976. He then spent two years as a postdoctoral Chaim Weizmann Fellow at MIT. In 1978, Phillips accepted an appointment as a physicist at the National Institute of Standards and Technology (NIST), a post he has held ever since.

In addition to his position at NIST, Phillips has been visiting professor of atomic physics at the Collège de France (1987), visiting scientist at the Ecole Normale Supérieure in Paris (1989–90), and adjunct professor of physics at the University of Maryland (1992–present). He was elected to the National Academy of Sciences in 1997.

Robert C. Richardson (1937–)

Along with David Lee and Douglas Osherhoff, Robert Richardson was named co-winner of the 1996 Nobel Prize in Physics for his studies of the superfluidity of helium-3. Helium-3 is an isotope of helium that undergoes dramatic changes in its physical properties at temperatures close to absolute zero (–273.15°C). Among these is the tendency for a liquid to behave in ways very different from those of a typical liquid, such as the tendency to flow upward along the sides of a container and to flow through very small apertures. These properties violate classical physical laws, but can be explained by quantum mechanical principles.

Richardson was born on 26 June 1937 in Washington, D.C. He received B.S. and M.S. degrees in physics from the Virginia Polytechnic Institute. His Ph.D. in physics was awarded by Duke University in 1966.

Richardson's first teaching appointment was at Cornell University in 1966, where he has remained ever since. He has served successively as research associate, assistant professor, associate professor, professor, and, most recently, F. R. Newman Professor of Physics. From 1990 to 1997, he was also director of the Laboratory of Atomic and Solid State Physics at Cornell.

Along with Lee and Osherhoff, Richardson was awarded the Simon Memorial Prize of the British Physical Society in 1976 and the Oliver E. Buckley Condensed Matter Physics Prize of the American Physical Society in 1981. He has also been elected a Fellow of the American Association for the Advancement of Science and the American Physical Society and is a member of the National Academy of Sciences.

Philip H. Scherrer (1946–)

Philip Scherrer was principal investigator for the Solar and Heliospheric Observatory (SOHO) Michelson Doppler Imager (MDI) that has provided new information about the process by which energy is transferred from the Sun's surface to its corona. The results of the SOHO MDI's discoveries about this process were announced in 1997.

Scherrer was born in California on 27 August 1946. He was awarded an A.B. and a Ph.D. in physics by the University of California at Berkeley in 1968 and 1973, respectively. After graduation, he served as research associate at the Institute for Plasma Research at Stanford University before joining the faculty at Stanford. He is currently professor in the Department of Physics and Center for Space Science and Astrophysics at Stanford.

Scherrer is, or has been, a member of more than a dozen professional organizations, including the SOHO Science Working Team (1987 to present), NASA's Helioseismology Steering Committee (1984–86), the Office of Naval Research's Space Science Committee (1992–93), and the National Research Council's Task Group on Ground-Based Solar Research (1997).

Richard E. Smalley (1943–)

Richard Smalley shared the 1996 Nobel Prize in Chemistry with Robert Curl and Harold Kroto for the discovery in 1985 of a new form of carbon known as buckminsterfullerene. Buckminsterfullerene molecules consist of 60 carbon atoms joined in a closed soccer-ball-like formation. Each molecule consists of 12 pentagons and some other number of hexagons. Buckminsterfullerene molecules are also known as "buckyballs." They and related molecules form the basis of a whole new category of materials known as nanotubes.

Smalley was born in Akron, Ohio, on 6 June 1943. He attended Hope College, in Holland, Michigan, from 1961 to 1963 and then transferred to the University of Michigan, from which he received a B.S. in chemistry in 1965. He then earned his M.A. and Ph.D. degrees from Princeton University in 1971 and 1973, respectively. In the period between 1965 and 1969, Smalley worked as a research chemist for the Shell Chemical Company.

Smalley's first academic appointment was at the University of Chicago, where he served from 1973 to 1976. He then moved to Rice University, where he has been assistant professor (1976–80), associate professor (1980–81), professor (1981–82), and Gene and Norman Hackerman Professor of Chemistry (1982–present). Since 1990, he has also held an appointment as professor of physics at Rice.

In addition to the Nobel Prize, Smalley has received the Irving Langmuir Prize in Chemical Physics from the American Physical Society (1991), the Popular Science Magazine Grand Award in Science & Technology (1991), the APS International Prize for New Materials (1992), the Ernest O. Lawrence Memorial Award from the U.S. Department of Energy (1992), the Welch Award in Chemistry from the Robert A. Welch Foundation (1992), the William H. Nichols Medal from the New York Section of the American Chemical Society (1993), the John Scott Award from the City of Philadelphia (1993), the Hewlett-Packard Europhysics Prize of the European Physical Society (1994), the Franklin Medal of the Franklin Institute (1996), and a Distinguished Public Service Award from the U.S. Department of the Navy (1997). He has also been granted honorary degrees by the University of Liege (Belgium) and the University of Chicago.

Paul Joseph Steinhardt (1952–)

Paul Steinhardt has made important contributions to the development of inflationary theory, the theory that describes the nature of the universe during the first few seconds following the Big Bang.

Steinhardt was born in Washington, D.C., on 15 December 1952. He attended the California Institute of Technology, from which he earned a B.S. degree in physics in 1974, and Harvard University, from which he received an M.A. and a Ph.D. in physics, in 1975 and 1978, respectively. He was appointed a junior fellow at Harvard for the period between 1978 and 1981 and then accepted an appointment at the University of Pennsylvania department of physics. He has served there as assistant professor (1981–83), associate professor (1983–86), professor (1986–89), and Mary Amanda Wood Chair Professor (1989–98). In 1998, he moved to Princeton University as professor of physics and associated faculty in the department of astrophysical sciences.

Steinhardt has served as visiting professor, scientist, or faculty at a number of institutions, including the Institute for Advanced Study at Princeton, New Jersey; the Institute for Theoretical Physics at Santa Barbara, California; the T. J. Watson Research Laboratory in Yorktown Heights, New York; the Johns Hopkins University; the Theory Group of the Stanford Linear Accelerator Center; and Tel-Aviv University in Israel.

Among his honors and awards have been the First Award of the Gravitational Research Foundation in 1993, a Freeman Dyson Fellowship of the Institute for Advanced Study in 1995, and a John Simon Guggenheim Fellowship in 1994–95.

Horst L. Störmer (1949–)

Horst Störmer was awarded a share of the 1998 Nobel Prize in Physics for his research on the fractional quantum Hall effect. This phenomenon involves the behavior of electrons moving through a material in the presence of a magnetic field. In 1982, Störmer and Daniel C. Tsui first observed a form of the Hall effect, discovered a century earlier, with potentially important applications to the development of semiconductor technology.

Störmer was born in Frankfurt am Main, Germany, in 1949. He earned a Ph.D. in physics at Stuttgart University in 1977. After leaving Stuttgart, he accepted an appointment at the Bell Laboratories of Lucent Technologies, a post he held for 20 years. In 1998, he was appointed professor of physics at Columbia University.

In addition to the Nobel Prize, Störmer has received the Oliver E. Buckley Prize from the American Physical Society, the Otto Klung Physics Award from Freie University in Berlin, and the Benjamin Franklin Medal in Physics from the Franklin Institute. He is a fellow of the American Physical Society and the American Academy of Arts and Sciences.

Daniel C. Tsui (1939–)

Daniel Tsui was honored with a share of the Nobel Prize in Physics in 1998 for his study of the fractional quantum Hall effect. The Hall effect was first observed in 1879 by Edwin Hall and has been the subject of intense research, especially over the past decade. Tsui and co-Nobelwinner Horst Störmer discovered an important variation of the Hall effect in 1982 that has the potential for important practical applications in semiconductor technology.

Tsui was born in Henan, China, in 1939. He earned a Ph.D. in physics at the University of Chicago in 1967. He was a research associate at the University of Chicago for one year before accepting a position as a member of the technical staff at Bell Laboratories in Murray Hill, New Jersey, in 1968. He held that post until 1982 when he was appointed professor in the department of electrical engineering at Princeton University. He is currently Arthur LeGrand Doty Professor of Electrical Engineering at Princeton.

Tsui has been awarded the 1984 Oliver E. Buckley Prize for Condensed Matter Physics and the 1998 Benjamin Franklin Medal in Physics, in addition to the Nobel Prize.

Carl E. Wieman (1951–)

Carl Wieman was one of the directors of a research team at the Joint Institute for Laboratory Astrophysics (JILA) in Boulder, Colorado, that first isolated a Bose-Einstein condensate consisting of about 2,000 atoms of rubidium at a temperature of 20×10^{-9} degrees kelvin. The discovery of the condensate provided confirmation for a theoretical speculation that had been offered six decades earlier by the Indian physicist Satyendra Bose and the Austrian-American physicist Albert Einstein.

Wieman was born in Corvallis, Oregon, on 26 March 1951. He attended the Massachusetts Institute of Technology, from which he received a B.S. in 1973, and Stanford University, from which he received a Ph.D. in 1977. His first academic appointments were at the University of Michigan, where he served as assistant research scientist (1977–79) and assistant professor of physics (1979–84). He then moved to the University of Colorado, where he was associate professor of physics from 1984 to 1987. In 1987, he was promoted to professor of physics at Colorado, a post he currently holds. Wieman was also appointed a Fellow of the Joint Institute for Laboratory Astrophysics (now JILA) in 1985, where he later served as chair from 1993 to 1995.

Among Wieman's many honors and awards are a Guggenheim Fellowship (1990–91), the E. O. Lawrence Award in Physics of the U.S. Department of Energy (1993), the Davisson-Germer Prize of the American Physical Society (1994), election to the National Academy of Sciences (1995), the Einstein Medal for Laser Science of the Society for Optical and Quantum Electronics (1995), the Fritz London Prize in Low Temperature Physics of the International Union of Pure and Applied Physics (1996), the Newcomb Cleveland Prize of the American Association for the Advancement of Science (1996), and the King Faisal International Prize for Science (1997). He was awarded an honorary doctorate of science by the University of Chicago in 1997. Wieman is the author or coauthor of more than 100 peer-reviewed papers.

Carl I. Wunsch (1941–)

Carl Wunsch, in collaboration with Walter Munk, proposed the concept of using sound waves to study the properties of the oceans in 1979. That idea has since led to the development of the Acoustic Thermometry of Ocean Climate (ATOC) project.

Wunsch was born in Brooklyn, New York, on 5 May 1941. He received his S.B. in 1962 and his Ph.D. in geophysics in 1966 from the Massachusetts Institute of Technology (MIT). Wunsch has taught and conducted research at MIT ever since, serving as assistant and associate professor (1967–75) and professor of oceanography (1975–76). Since 1976, he has been Cecil and Ida Green Professor of Physical Oceanography.

Wunsch has also served as visiting professor at Harvard University (1980) and the University of Washington (1980), and as visiting scientist or visiting fellow at the University of Cambridge (1969, 1974–75, 1981–82), Princeton University (1993–94), Groupe de Recherche de Geodesie Spatiale–CNES/CNRS, Toulouse, France (1994), and the Jet Propulsion Laboratory (1994–present).

Arjun G. Yodh (1959–)

Arjun Yodh's special fields of interest include aspects of chemical, condensed-matter, and optical physics. In the early 1990s, he was involved in the study of certain phenomena that appeared to violate the principle of entropy. In his work with graduate student Peter Kaplan, Yodh studied the formation of discrete crystalline structures within chaotic systems.

Yodh was born in Pittsburgh on 7 July 1959. He earned a B.Sc. from Cornell University in 1981, and M.S. and Ph.D. degrees from Harvard University in 1982 and 1986, respectively. He spent the period from 1986 to 1988 as a postdoctoral research associate at the AT&T Bell Laboratories. He was then appointed assistant professor of physics at the University of Pennsylvania in 1988. In 1993, he was promoted to the post of associate professor and, in 1997, to the positions of professor of radiation oncology and professor of physics and astronomy at the University of Pennsylvania. Since 1997, he has also been adjunct professor in the School of Bioengineering at Drexel University.

Yodh's research activities have been supported by the Lilly Foundation (1988–89), the National Science Foundation's Presidential Young Investigator Program (1990–95), the Alfred P. Sloan Foundation (1991–94), and the Office of Naval Research (1991–94). He currently holds the William Smith Term Chair (1997–2002) at the University of Pennsylvania.

Anton Zeilinger (1945–)

Anton Zeilinger led a research group at the University of Innsbruck that reported in 1997 on the teleportation of elementary particles. Although the breakthrough has no obvious practical applications, the accomplishment has enormous theoretical consequences because it confirms a long-held theoretical assumption that particles could be transported instantaneously across space.

Zeilinger was born in Ried/Innkreis, Austria, on 20 May 1945. He received a Ph.D. in physics in 1971 from the University of Vienna and then accepted a position as Universitätsassistent at the Atominstitut in Vienna. He left Vienna in 1981 to serve two years as associate professor of physics at the Massachusetts Institute of Technology (MIT). He then returned to Vienna where he took a position as Ausserordentlicher at the Technical University of Vienna. In 1990, he was promoted to Ordentlicher Universitätsprofessor (full professor) of experimental physics at the University of Innsbruck. In 1999, he was appointed chair of experimental physics at the University of Vienna.

Zeilinger has also served as visiting professor or adjunct professor at the University of Melbourne, in Australia; Hampshire College; Collège de France, in Paris; Merton College, Oxford University; and the Los Alamos National Laboratory.

Zeilinger has been honored with many special awards, including the Prix "Vinci d'Excellence" of the LVHM Foundation in Paris and the European Optics Prize of the European Optical Society. In 1996, he was named Austrian Scientist of the Year.

Fritz Zwicky (1898–1974)

Fritz Zwicky's early fame arose because of his research on supernovas and his discovery of how these objects differ from ordinary novas. In recent years, however, his investigations of the mass of the universe has gained even more attention. Zwicky attempted to calculate the mass of the universe based on stars and other objects whose mass could be determined based on their luminosity. His discovery of an enormous mass difference between calculations and experimental observations has led to the concept of "dark matter," matter that is not visible but that apparently must be present in the universe.

Zwicky was born in Varna, Bulgaria, of Swiss parents, on 14 February 1898. He attended the Federal Institute of Technology in Zürich, Switzerland, from which he received his doctorate in 1922. Three years later, Zwicky emigrated to the United States, where he took a position on the faculty at the California Institute of Technology (Caltech). He also joined the staff at the Mount Wilson and Palomar observatories. Zwicky spent the rest of his professional career at these two institutions. He died in Pasadena on 8 February 1974.

CHAPTER FIVE
Social Issues in Physics

HEALTH EFFECTS OF ELECTROMAGNETIC FIELDS

W ell, do they or don't they? Electromagnetic fields, that is. Do the electromagnetic fields (EMFs) that surround electrical transmission wires, electrical machinery, and electrical devices in the home cause health problems for humans or not? Scientists have been wrestling with that problem for more than two decades now. And the answer still seems unclear.

This question first arose more than 20 years ago in what was then the Soviet Union. A number of men and women who worked around high voltage electrical equipment began to complain of fatigue, headaches, appetite loss, insomnia, and reduced sexual drive. The question arose as to whether exposure to EMFs was responsible for these health problems.

Studies conducted by Soviet scientists found no evidence that EMFs were responsible for the workers' problems. However, it was not long before similar complaints were being heard in other parts of the world, especially in the United States and Great Britain. People began to seek medical advice, to register their complaints with power regulatory agencies, and even to take violent action against electric power companies themselves.

In one case, judges in New York State granted approval for the construction of two new power lines about which nearby residents had complained. But the judges also acknowledged that evidence about possible health effects from the lines was too strong to ignore completely. In another case, people living near a proposed new power line in Minnesota and North Dakota tried to sabotage the project. They caused more than $9 million in damage before the line was actually completed in 1978.

At first, most scientists did not take very seriously the idea that the kind of EMFs to which people are routinely exposed could affect the health of humans and other animals. No existing theory or body of research existed to support that idea. But, gradually, more and more researchers became interested in this issue. They carried out two general types of studies: laboratory experiments and epidemiological studies.

In the laboratory experiments, cells, tissues, and small animals were exposed to EMFs and then studied for damage. In epidemiological studies, researchers compared the health histories of people living or working close to, and more distant from, sources of EMFs to see what differences they could find.

Over time, a body of evidence began to develop suggesting that EMFs may have harmful effects on cells, tissues, and animals. In 1979, for example, two scientists reported that children living near high voltage power lines are more likely to develop cancer than those who do not.

By the end of the 1990s, dozens of studies had been completed on this question. In fact, a bibliography on electromagnetism and health compiled by Richard W. Woodley for the Bridlewood Residents Hydro Line Committee included 754 items on the subject. Overall, the results of these studies were largely inconclusive. They often showed a correlation between exposure to electromagnetic fields and cancer rates or reproductive problems. But scientists know that a high correlation does not necessarily mean a cause-and-effect relationship. The fact that people who live near power lines have a higher-than-average rate of cancer does not automatically mean that the power lines *caused* the cancers. Some third factor, of which scientists are not aware, may also be involved.

Other studies showed no correlation between EMFs and health conditions. They suggested that people who live near high intensity electromagnetic sources are as healthy as those who do not.

The fundamental problem, of course, is that health effects caused by EMFs are likely to be very small and to be cumulative over many years. It is extremely difficult to say whether a 50-year-old woman has developed breast cancer because of living near a power transmission line for 10 years, or whether other factors are involved in the disease.

Between 1996 and 1998, three major studies of the health effects of EMFs were reported. They were (1) a review of the literature by a research team from the National Research Council of the National Academy of Sciences in 1996, (2) the National Cancer Institute's Linet Study of 1997, and (3) the Royal Adelaide (Australia) Hospital ELF Mice Study of 1998. All three of these studies concluded that there was no basis for concern about the health effects of EMFs on humans or other organisms.

Nevertheless, these studies appear not to have calmed scientific and public concerns about the health effects of EMFs. A number of researchers are still convinced that specific cell and molecular damage can be traced to exposure to electromagnetic fields, and they have developed a number of theories and designed new experiments to test these ideas.

For example, some scientists believe that EMFs can damage a chemical known as melatonin that occurs naturally in the human body. Melatonin is a powerful anticarcinogenic agent that has been found to be very effective in suppressing the development of cancer cells both in test tubes and in the body. Studies are now underway to understand how EMFs affect the production of melatonin, which, in turn, may be related to the growth of cancer cells in the body.

A report issued in late 1998 by an expert panel sponsored by the National Institute of Environmental Health Sciences and the U.S. Department of Energy reflected the current state of confusion over the effects of EMFs on human health. That report concluded that data on the effects of EMFs in causing various forms of cancer were inconclusive. But it also acknowledged that laboratory studies did suggest that EMFs were potential carcinogenic agents. Finally, they recognized that health problems less severe than cancer, such as sleep disorders and drug inhibitions, might be traceable to the effects of EMFs.

The debate over the health effects of electromagnetic fields took an unexpected turn in July 1999 when a researcher was accused of falsifying data on this subject. The researcher, employed at the Lawrence Berkeley National Laboratory, was charged with deliberately changing his data so that it would show biological effects of EMF on the way in which cells use calcium. The researcher has denied these charges, and it is not clear how false results in two out of more than a thousand studies could or will affect this debate in the future.

After nearly 30 years of research, then, the jury is still out on the question posed at the beginning of this section. Most scientists involved in EMF studies would probably agree that electromagnetic fields are not a major cause of breast cancer, brain cancer, leukemia, and other serious health problems. However, EMFs may well be a factor that contribute to the development of such diseases, as well as to less serious health problems.

References

Beardsley, Tim, "Guessing Game," *Scientific American*, March 1991: 30–32.

Brodeur, Paul, "The Annals of Radiation," *The New Yorker*, 12 June, 19 June, and 26 June 1989: 51–88, 47–77, and 39–68, respectively.

"The Electromagnetic Radiation Health Threat, Parts I and II" <http://www.nzine.co.nz/features/neilcherry.html> accessed 30 April 1999.

"Jury Still Out on EMFs and Cancer," *Science News*, 22 August 1998: 127.

LaMacchia, Diane, "Study of Biological Effects of Electromagnetic Radiation Inconclusive," <http://www.lbl.gov/LBL-Science-Articles/Archive/electromagnetic-radiation-study.html> accessed 30 April 1999.

Maisch, D., and B. Rapley, "Powerline Frequency Electromagnetic Fields and Human Health—Is It the Time to End Further Research?" <http://www.tassie.net.au/emfacts/3studies.html> accessed 30 April 1999.

Marshall, Jonathan, "Cell Phones Linked to Lymphoma in Mice," *San Francisco Chronicle*, 9 May 1997: A1.

Noland, David, "Power Play," *Discover*, December 1989: 62–68.

Overbeck, Wayne, "Electromagnetic Fields and Your Health," <http://www.arrl.org/news/rfsafety/wo9404.html> accessed 30 April 1999.

Raloff, Janet, "EMFs' Biological Influences," *Science News*, 10 January 1998: 29–31.

United States Congress, Office of Technology Assessment, *Biological Effects of Power Frequency Electric and Magnetic Fields: Background Paper*, June 1989.

Woodley, Richard W., "Bibliography on Electromagnetic Radiation and Health," <http://www.envirolink.org/seel/emf.html> accessed 30 April 1999.

THE INTERNATIONAL SPACE STATION

"Space: The Final Frontier." That slogan launched a successful series of television programs and movies. For well over three decades, it has also been the rallying cry for a large body of scientists, politicians, and others who see space as the next great step forward in humankind's effort to understand the world around us. The next stage in that effort is scheduled to be the International Space Station.

Space Station Design

The International Space Station (ISS) is planned to be a structure consisting of laboratories, living space, a power plant, escape devices, and other support components in which gravity-free experiments can be carried out. It is also intended to serve as a vehicle from which trips to Mars can be launched.

ISS is truly an international venture, with 16 nations participating in its design, construction, and operation. The two lead nations of the ISS consortium are those with the greatest experience in space research: the

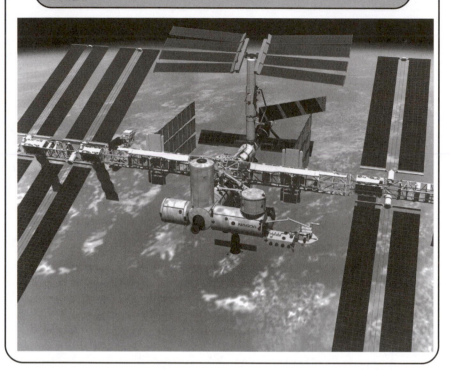

Artist's concept of the International Space Station, as it will appear when its assembly sequence is complete. *Courtesy of NASA.*

United States and Russia. Construction of the ISS was originally planned to take place over a 55-month period during which components would be carried into orbit on 73 separate rocket launches. The first launch was planned for 1997, and the station was scheduled to be ready for occupancy in 2002.

The ISS has been a "hard sell" in the United States. The National Aeronautics and Space Administration (NASA) has made every effort to point out the potential practical benefits of the station in everyday life, benefits such as a better understanding of the ways in which the human body functions and the development of new materials and new manufacturing processes. NASA has also argued that the ISS will open a new era of international cooperation among nations around the world and has forecast that the ISS will become a beacon of hope and a symbol of scientific progress for the next generation of Americans. These arguments are, in some regards, similar to those that have traditionally encouraged support for space programs in the U.S. Congress for more than four decades.

Objections to the Space Station

Support for the ISS has not, by any means, been unanimous. One problem is that, for all NASA's efforts to emphasize other aspects, the space station is fundamentally concerned with basic research. Its basic goal is simply to learn more about outer space. Whatever practical benefits may derive from the station will probably be secondary to this fundamental effort to learn more about the universe in which we live. Legislators have always been torn between the desire to provide funding for experiments that push scientific frontiers even further into space and the need to allocate money for the solution of the practical everyday problems of ordinary Americans.

It is hardly surprising, then, that requests for ISS funding have met with vigorous debates in Congress and, on occasion, very close votes over budgetary decisions. In 1993, for example, Representatives Tim Roemer (D-IN) and Dick Zimmer (R-NJ) joined together to offer an amendment to the NASA authorization bill calling for a cancellation of the space station. That amendment failed by a single vote, 215 to 216. Since that time, other amendments also have been proposed to various bills calling for an end to the space station program. All have failed, usually by much larger margins.

Technical Problems

Work on the ISS has been hampered not only by political controversy, but also by a series of technical problems. For example, the initial launch of the first station components scheduled for 1997 had to be canceled when engineers at the Boeing Company discovered in 1996 that connecting nodes joining sections of the station had been improperly designed. The nodes had to be redesigned and rebuilt at an additional cost of $100 million and a delay of six months in the overall project.

Then, in 1997, the Russian Space Agency announced that it would not be able to complete on time the service module in which crews would work and operate the station. The Russian government had provided only one-tenth of the money promised and needed. As a result, launch of the station's first unit, the Russian-built tug, was delayed a year.

By the end of 1998, there was even more bad news for the ISS. A study commissioned by NASA concluded that unexpected technical and economic problems had caused a three-year delay in completing the project. Furthermore, the cost of the station was predicted to be closer to $24.8 billion than to the $17.4 billion originally estimated. One problem was that Russia was experiencing both economic and technical problems in completing its part of the project. As a result, Russian involvement in the station was actually going to increase the cost for other participants, rather than saving them money, as had originally been anticipated.

Nevertheless, ISS supporters felt confident that the station would still be completed, if not on time, at least shortly thereafter. However, critics and members of Congress appeared to be running out of patience. As Senator Barbara Mikulski (D-MD) said, "There's always one more promise." The question troubling NASA and ISS supporters was how long they could continue to ask for funding when those promises were not being kept.

References
Davis, Brett, "Russia's Delay in Space Station Project Worries NASA," *The Oregonian*, 8 March 1997: A15.

Dunn, Marcia, "Space Station Promise Is Hard to Keep," *San Francisco Chronicle*, 14 September 1996: A3.

Eisler, Peter, "House Approves NASA Plan to Go Ahead with Space Station," Gannett News Service, 29 June 1994; also <http://www.elibrary.com/id/192/192/get . . . docid=90977 . . . > accessed 27 June 1998.

"International Space Station: It's about Life on Earth . . . and Beyond" <http://spaceflight.nasa.gov/station/reference/factbook/beyond.html> accessed 31 May 1999.

Lawler, Andrew, "Report Sees More Overruns, Delays," *Science*, 1 May 1998: 666.

"Space Legislation: House Votes" <http://www.ari.net/nss/voters_guide/page8.html> accessed 30 April 1999.

SETI

The ultimate experiment in physics may well be SETI. SETI is the acronym for the Search for Extraterrestrial Intelligence.

Humans have long believed that we are not the only intelligent creatures in the universe. Until recently, however, speculations about the existence of intelligent life beyond the Earth have been restricted to the realm of science fiction. About 50 years ago, however, it became possible to design more scientific and less fictional methods for seeking out intelligent life in the universe.

Current interest in SETI has been spurred by developments in two fields. First, there has been a dramatic expansion and improvement in the tools available to astronomers who wish to explore the regions beyond our Solar System. These new technologies include powerful telescopes that can detect radiation in the whole range of the electromagnetic spectrum, such as radio telescopes, X-ray telescopes, and gamma-ray telescopes.

Second, SETI enthusiasts have been encouraged by the discovery of a number of planets and planet-like objects outside of our own Solar

System. Although it is too early to tell, the potential for intelligent life certainly exists on such planet-like objects. In any case, the mathematical probability of finding intelligent life somewhere in the vast reaches of the universe appears to be near certainty. Some authorities have estimated that as many as 10 billion Earth-like planets exist in outer space.

SETI research has taken primarily two forms. First, it has employed some of the largest telescopes ever built to focus on various regions of space with the objective of picking up signals that might be sent to us by intelligent beings elsewhere. Second, it has developed a communications program intended to convey to such beings something about our own world. Two gold plaques containing a message from Earth scientists to intelligent beings outside the Solar System were included, for example, in the Pioneer 10 and 11 space probes launched from Earth in 1972 and 1973, respectively. Those plaques were inscribed with mathematical symbols designed to communicate the nature of our planet and the beings who inhabit it.

In the years following World War II, SETI research was financed primarily by individual research teams and institutions interested in the subject. Scientists scavenged whatever time they could on telescopes to listen for signals from outer space or to transmit their own messages to potential listeners on other worlds.

Then, in 1991, the federal government made its first major commitment to SETI research. The U.S. Congress voted to give $100 million to the National Space and Aeronautics Administration (NASA) for the support of a 10-year project for the search for intelligent life in the universe. Part of that money was to be channeled through the SETI Institute, located near Stanford University. The SETI Institute was created by Frank Drake, one of the "grand old men" of SETI.

This commitment turned out to be short-lived. After only two years of the NASA project, Congress became convinced that it had wasted taxpayers' money. No results had been produced from the SETI research, and critics argued that the government was throwing money away on a fanciful scheme. They made a convincing argument that SETI money could be better spent on important social programs, such as improved health care and education.

Such arguments are familiar, of course, in discussions over basic research projects like SETI. Those who argue for such projects can only point to fundamental progress in our understanding of the natural world. They cannot claim that cures for disease, new commercial products, or other tangible improvements will result from their research. They can only hope that those who hold the purse strings will see some value in pushing forward the limits of human knowledge.

In 1993, the U.S. Congress had not reached that point in their thinking. They canceled the 1991 NASA project and decided to provide no further funds for SETI research.

As with many government decisions, Congress's action in 1993 turned out to be revocable. In fact, by the end of the 1990s, there was evidence that NASA was once again interested in promoting SETI research. Dan Goldin, administrator of NASA, announced plans for a new institute of astrobiology within NASA, increased support for NASA's Origins Program (which deals with the origin and distribution of life in the universe), and plans to construct a telescope to orbit beyond Jupiter and search for Earth-like planets outside the solar system.

In the meanwhile, private funding for SETI research has continued. The SETI Institute, for example, continues to sponsor research, largely through the financial support of David Packard and William Hewlett, cofounders of the Hewlett-Packard Company; Gordon Moore, chairman of the Intel Corporation; and Paul Allen, cofounder of Microsoft, Inc.

In addition to the SETI Institute, research on extraterrestrial intelligence is being conducted at four other locations. They are Project Phoenix, in Palo Alto, California; Project Serendip, sponsored by the University of California at Berkeley; Projects Meta and Beta, sponsored by the Planetary Society and Harvard University; and Project Big Ear, sponsored by Ohio State University.

For the foreseeable future, however, SETI research will be surrounded by questions about the use of taxpayer money for "far-out" basic research that may never have any practical application. Such questions are similar to those that arise over the funding of the International Space Station, the construction of large particle accelerators, and other very expensive projects in basic research.

References

Achenback, Joel, "Search for Aliens No Longer on Edge of Science Galaxy," *The Sunday Oregonian*, 15 March 1998: G3.

Boyd, Robert S., "Is Anyone Out There?" *The Oregonian*, 23 May 1996: B11.

Bracewell, Ronald N., *Intelligent Life in Outer Space*, San Francisco: W. H. Freeman and Company, 1975.

Davidson, Keay, "Search for Alien Life Forms Expands," *The San Francisco Examiner*, 12 April 1998: D1.

Sagan, Carl, *The Cosmic Connection: An Extraterrestrial Perspective*, New York: Dell Publishing, 1973.

Shklovskii, I. S., and Carl Sagan, *Intelligent Life in the Universe*, San Francisco: Holden-Day, Inc., 1966.

NUCLEAR WASTES

Why aren't there more nuclear power plants in the United States? Many people feel that nuclear power is the cleanest, safest, most economical method for generating electrical power. In other countries of the world, that argument has been persuasive. In France, for example, more than half of the electrical power generated comes from nuclear power plants.

In the United States and other nations, however, strong opposition to the construction of nuclear power plants has developed. Two major factors are responsible for this trend. First, there have been serious concerns about the safety of such plants. Disasters such as those that occurred in 1979 at the Three-Mile Island Plant in Pennsylvania and in 1986 at the V. I. Lenin Atomic Power Station at Chernobyl, Ukraine, have convinced many observers that nuclear power is simply too dangerous to be used on a large-scale basis.

The second factor inhibiting the growth of nuclear power has been the problem of nuclear wastes. Nuclear wastes are radioactive materials left over from, or produced during, various processes. These processes include not only the generation of electricity in nuclear power plants, but also the manufacture of nuclear weapons and the use of nuclear materials in medical research, diagnosis, and treatment.

Most power companies and government agencies seem not to have given much thought to the problem of nuclear wastes in the early history of the use of nuclear power. The 1950s and 1960s were heady times for nuclear power enthusiasts who saw this new technology as the solution to many energy-related problems. The behavior of industry and government seemed to suggest that a solution to waste disposal problems would eventually be found. Meanwhile, wastes were simply stored in much the same way as other industrial wastes: They were buried underground or dumped into deep lakes and the oceans. In some cases, they were simply covered up and stored on vacant lots outside the plants where they were generated.

An example of the problems that have developed as a result of this philosophy is a cache of 46,000 metal cylinders currently being stored in Paducah, Kentucky; in Oak Ridge, Tennessee; and in Piketon, Ohio. The cylinders weigh up to 14 tons each and have been slowly rusting away since the 1940s. They contain highly radioactive uranium and other elements, as well as toxic hydrogen fluoride gas. The U.S. Department of Energy announced plans in 1998 to offer these cylinders for sale to potential recyclers of these materials. Many authorities, however, doubt that anyone will be interested in that offer.

The Nuclear Waste Policy Act of 1982

The problem has been that no one—industry or the federal government—had any systematic plan for the disposal of nuclear wastes for over three decades after those wastes began to accumulate in the 1940s. Finally, in 1982, the U.S. Congress passed the Nuclear Waste Policy Act. That act directed the Department of Energy (DOE) to develop a plan for storing spent (used) fuel rods removed from nuclear power plants. As of 1997, more than 32,000 tons of such wastes had accumulated and were being stored close to the reactors from which they were removed or at other "temporary" sites.

After an extensive survey, DOE selected Yucca Mountain, Nevada, as a site for burial of these wastes, which are designated high-level wastes because of the intense, long-lived radiation they produce. Yucca Mountain is located on the Nellis Air Force Base, the Nevada Nuclear Test Site, and land managed by the federal Bureau of Land Management.

The DOE plan called for construction of more than 130 miles of tunnels buried 1,000 feet underground inside the mountain. Wastes would be hauled into the tunnels by railroad cars and deposited in side tunnels leading off the main tunnel. The side tunnels would then be sealed up, isolating the wastes for hundreds of years. According to this plan, up to 77,000 tons of wastes could be accommodated, a capacity that would solve the nation's nuclear waste disposal for high-level wastes for 50 years.

The DOE plan ran into opposition as soon as it was proposed. As expected, residents of Nevada and politicians in the state objected to the use of their state as a "nuclear graveyard." Environmentalists worried about the escape of radioactive wastes into the ground and, then, into groundwater. They pointed out that the wastes would continue to pose a health and environmental threat for thousands of years into the future. No one knew, they said, how natural forces might cause movement or exposure of the wastes.

Because of this opposition and technical problems that developed, progress at Yucca Mountain moved forward only very slowly. Originally scheduled to open in 1998, the facility was less than 5 percent completed on January 1 of that year. Opening date had by then been moved to 2010. The possibility of even further delays beyond that point, however, were still possible.

In the meantime, power companies had lost patience with the federal government. The original Nuclear Waste Policy Act had included a tax of 0.1 cent on each kilowatt hour of electricity generated by nuclear power in the United States. Power companies had been collecting that tax for the better part of two decades and had contributed more than $4 billion for development of a nuclear waste disposal site. Yet the construction of the Yucca Mountain facility had barely begun. At the same time, power companies were still spending tens of millions of dollars annually to store nuclear wastes at other sites.

Having lost patience, a group of 25 utilities and 18 state commissions filed suit against DOE in 1996 for its lack of action. In July of that year, a federal appeals court ruled that DOE was legally responsible for nuclear waste, "ready or not," on 31 January 1998. If Yucca Mountain was *not* ready (as it turned out not to be), DOE was potentially responsible for tens of billions of dollars in refunds and storage costs to utilities and states. The ongoing drama of nuclear waste disposal was not only not over . . . it had apparently just begun!

Storage of Low-Level Wastes in New Mexico

Plans for dealing with low-level nuclear wastes have had an even longer history than those for high-level wastes. Low-level radioactive wastes include gloves, tools, and other materials used in working with plutonium and other so-called transuranic elements (elements heavier than uranium).

As early as the 1970s, DOE had begun plans for the construction of a waste disposal site for low-level wastes in southeastern New Mexico. The site was called the Waste Isolation Pilot Plant (WIPP). WIPP was to consist of a huge burial vault in salt beds 2,150 feet below the Earth's surface. Construction of the site was begun in the late 1980s and completed in 1988. Total cost of the project was $1.8 billion.

In comparison with the Yucca Mountain plan, WIPP appeared to move forward with a certain level of efficiency. It was ready to begin receiving wastes almost on time and nearly within budget, given the usual trend toward overruns in government projects.

One "fly in the ointment" remained, however. WIPP was not allowed to open until DOE wrote an environmental impact statement acceptable to the U.S. Environmental Protection Agency (EPA). That requirement turned out to be much more difficult than anyone had anticipated. In fact, 10 years after DOE announced that the site was ready to receive wastes, the necessary environmental approval had still not been granted. One of the major concerns expressed by the EPA was that radioactive materials might escape into the salt dome and, from there, into underground water systems. Critics of WIPP also pointed out that DOE had already developed an even more secure burial site in Nevada. Why couldn't Yucca Mountain also be used for the storage of low-level wastes? they asked.

DOE officials argued that neither of these arguments was valid. First, the New Mexico salt dome was laid down during the Permian Period 225 million years ago and has remained stable ever since. Further, the Yucca Mountain facility was not designed with the capacity needed to store low-level as well as high-level wastes.

Resolution of the New Mexico site debate was finally achieved in May 1998 when the EPA announced its approval of the DOE site plan. The residents in areas surrounding the site were ecstatic and began making plans to greet the first shipment of radioactive wastes on 19 June 1998. Over the planned life of the site, an additional 37,000 shipments would arrive over a 30-year period from "temporary" holding sites in California, Colorado, Idaho, Illinois, Nevada, New Mexico, Ohio, Tennessee, South Carolina, and Washington.

References

Barlett, Donald L., and James B. Steele, *Forevermore: Nuclear Waste in America*, New York: W. W. Norton, 1985.

Brooke, James, "Underground Haven or a Nuclear Hazard? *New York Times*, 6 February 1997: A12.

Long, James, and Jim Barnett, "Warehousing Nuclear Waste," *The Oregonian*, 12 March 1997: A12.

"No Quiet Burial for Nation's Nuclear Waste Debate," *San Francisco Chronicle*, 21 January 1997: A5.

Sahagun, Louis, "N.M. City Embraces Radioactive Waste Plant, *The Oregonian*, 9 June 1998: A12.

Turk, Jonathan, and Amos Turk, *Environmental Science*, 4th edition, Philadelphia: Saunders College Publishing, 1990: 399–403.

Wald, Matthew L., "Plan to Eliminate Nuclear Waste Has Hard Time Finding a Market," *San Francisco Chronicle*, 17 March 1998: A7.

Warrick, Joby, "N.M. Nuclear Waste Dump Gains License to Operate," *The Oregonian*, 14 May 1998: A9.

NUCLEAR WEAPONS

The most significant social issue relating to physics may well be that of nuclear weapons. During the 1940s, researchers discovered methods by which two reactions, nuclear fission and nuclear fusion, could be harnessed to manufacture weapons. In *nuclear fission*, neutrons are used to break apart atoms of uranium or plutonium. In *nuclear fusion*, atoms of hydrogen are forced together to make atoms of helium. In both reactions, very large amounts of energy are produced. Fission and fusion weapons are, by far, the most powerful weapons ever invented by humans.

Over the past half century, a number of nations have developed the technology for making nuclear weapons. Given the level of animosity among nations and the number of local wars that have grown up over that period of time, it is somewhat remarkable that only two nuclear weapons have ever been exploded in a conflict: the two fission ("atomic") bombs dropped on Hiroshima and Nagasaki, Japan, at the end of World War II.

From the moment nuclear weapons were first conceived, there has been debate as to whether they should ever be built or used. During research on nuclear weapons, for example, some of the nuclear scientists themselves argued that the weapons should never actually be built or used. They understood the level of destruction they would produce and felt that no justification could be made for the widespread, intentional devastation they would cause. Those scientists' views did not prevail, of course, but the controversy over nuclear weapons quickly spread to the general public when reports of the Hiroshima and Nagasaki bombings were released.

As appalling as nuclear weapons may seem, they have many defenders. Any nation that possesses a nuclear weapons capability holds a potential trump card in any conflict in which it becomes involved. The United States, as the most extreme example, certainly owes at least some share of its claim to being the leading nation in the world to its nuclear weapons capability.

Nuclear Weapons Treaties

It is no small wonder, then, that efforts to reach international agreement on the control of nuclear weapons has moved forward so slowly. In fact, progress in this area has traditionally been measured in very small steps that take years of negotiation to accomplish. Listed below are some of the agreements that have been reached about the control of nuclear weapons.

The *Limited Test Ban Treaty of 1963* prohibits the testing of nuclear weapons in the atmosphere, outer space, or under water. It was originally negotiated by the United States, the United Kingdom, and the Soviet Union. By 1998, 116 nations had signed the treaty. Two nations with nuclear capability have never signed it: France and China.

The *Nuclear Non-Proliferation Treaty of 1968* was a temporary agreement designed to prevent the spread of nuclear weapons. It was also designed to prevent nations from diverting nuclear materials from peaceful applications to weapons development. Three major nuclear powers, the United States, the United Kingdom, and the Soviet Union, along with 133 nations without nuclear weapons, eventually signed the treaty. In May 1995, the treaty was made permanent by action of the United Nations.

The *Anti-ballistic Missile Treaty of 1972* was a bilateral agreement between the United States and the Soviet Union. It placed limitations on the areas in which ABMs (anti-ballistic missiles) can be deployed and prohibited the development of many critical features of ABM technology.

The *Strategic Arms Limitation Treaty (SALT) I of 1972* was also an agreement reached between the United States and the Soviet Union. It was given a limited lifetime—five years. After that period, it had to be renegotiated by the two nations. The major component of SALT I was a freeze on the number of strategic ballistic missiles to be developed by the two nations.

The *Strategic Arms Limitation Treaty (SALT) II of 1979* was negotiated as a replacement for SALT I, which expired in 1977. The provisions of SALT II mirrored those of SALT I in many ways and provided further limitations on the production of strategic nuclear weapons.

The *Strategic Arms Reduction Treaty (START) I of 1991* was also a bilateral treaty between the United States and the Soviet Union. It was designed to reduce—beyond the provisions of SALT II—the number of nuclear weapons that could be developed by either nation.

The *Strategic Arms Reduction Treaty (START) II of 1993* was a bilateral agreement between the United States and Russia. It was signed by Presidents Bush and Yeltsin in January 1993 and further reduced the number of nuclear weapons to be developed by either nation.

The *Comprehensive Test Ban Treaty (CTBT) of 1996* was finally approved in 1996 by the United Nations after having been discussed by that body for more than four decades. The treaty was signed by 149 nations. As its title suggests, this treaty is intended to drastically curtail the development and testing of nuclear weapons by all countries of the world.

Recent Developments

As with all international treaties, those relating to nuclear weapons must pass through two steps of approval. The first step is simply the signing of the treaty, an act in which those individuals involved in negotiating the treaty pledge their support for the decision by signing the treaty document. The dates associated with the treaties listed above indicate the year in which those treaties were signed.

However, treaties must also be ratified, or confirmed, by the legislative or other decision-making bodies of the nations involved. In the United States, for example, treaties have to be approved by the U.S. Senate. No treaty is considered to be finally approved until it has been ratified. In many cases, ratification takes years. For example, it took the U.S. Senate more than three years to ratify the START II Treaty.

A similar pattern has held true for the most recent agreement, the Comprehensive Test Ban Treaty (CTBT). The delay in approving the

CTBT illustrates the power of individual senators to block agreements with which they do not agree. In this case, it has been Senator Jesse Helms (R-NC), chairman of the Foreign Relations Committee, who has held up action on the CTBT. Helms has argued that the CTBT is not in the best interest of the United States and that it should not be ratified at all. In his position as chair of the Foreign Relations Committee, he is able to impose that view on any action, or lack of action, that the Senate may take on the treaty.

As eager as they are to have the CTBT ratified, proponents of nuclear weapons limitations have at least had some reason to be optimistic about their case. For a number of years, major nuclear powers, such as the United States and Russia, have exercised admirable restraint in the testing of nuclear weapons. However, that modest cause for optimism was shattered in 1998 when two "junior" nuclear powers, India and Pakistan, both carried out tests of nuclear weapons.

The impetus for these tests appeared to be nothing other than the continued and passionate animosity between these two nations. The Indian tests, which occurred between May 11 and 13, were apparently designed to convince Pakistan that India would be able to protect itself in case its traditional enemy decided to attack. Some observers believe that the tests were also intended to encourage a sense of pride and nationalism among the people of India.

Pakistan chose not to ignore the show of Indian bravado and launched its own nuclear tests two weeks later. Pakistani officials argued that they had no choice but to demonstrate that they, too, had a nuclear weapons capability. Indeed, their own tests may have helped to create a certain level of stability in the fragile stalemate between the two nations because of their mutual reluctance to initiate a nuclear conflict.

In any case, other nations around the world were horrified at the prospect of a continued proliferation of nuclear weapons by more and more nations. It seemed increasingly important that some international agreement, whether it be the CTBT or a comparable treaty, be reached to limit the further spread of such weapons.

References

"Arms Control Treaties" <http://www.atomicarchive.com/ACTreaty.shtml> accessed 30 May 1999.

"Background Information" (on START II ratification) <http://infomanage.com/nonproliferation/treaties/startii.html> accessed 30 May 1999.

Bagla, Pallava, and Andrew Lawler, "Experts Search for Details after Indian Nuclear Tests," *Science*, 22 May 1998: 1189.

Gordon, Michael R., "U.S. to Offer Deeper Nuclear Cuts in Start III Talks," *San Francisco Examiner*, 9 March 1997: A7.

Jensen, Holger, "Test Ban Treaty on the Shelf," <http://www.nando.net/newsroom/041198/voices6_15590_noframes.html> accessed 30 May 1999.

Sagan, Scott Douglas, and Kenneth Neal Waltz, *The Spread of Nuclear Weapons: A Debate*, New York: W. W. Norton, 1995.

Schroeer, Dietrich, *Science, Technology and the Nuclear Arms Race*, New York: John Wiley and Sons, 1984.

Turner, Stanfield, *Caging the Nuclear Genie: An American Challenge for Global Security*, Boulder, CO: Westview Press, 1997.

"U.N. Passes Nuclear Test Ban Treaty," *San Francisco Chronicle*, 11 September 1996: A1.

IRRADIATION OF FOOD

The word *irradiation* has unpleasant, even horrifying, implications for many people. Particularly for those familiar with the effects of nuclear weapons, the term conjures up visions of people being bombarded with nuclear radiation that will kill them, at worst, or at least make them extremely ill with "radiation sickness." Even those for whom medical radiation treatments have brought relief from cancer and other serious illnesses are aware of the terrible temporary effects of exposure to radiation.

Objections to Food Irradiation

Thus, it is hardly surprising that so much opposition would be expressed to the use of radiation in the treatment of foods to prevent spoilage. On an emotional level, many people are probably concerned that the use of nuclear radiation to preserve foods is likely to cause health problems similar to those that result from nuclear weapons and medical radiation treatments. They worry that traces of radiation may remain in irradiated foods, potentially causing radiation health problems to those who eat such foods.

But beyond that "gut" level of concern, many people have more reasoned objections to the irradiation of food. They fear that radiation may change the molecular composition of foods in such a way as to produce new compounds that are deleterious to human health. They worry that such compounds may have long-term effects, such as the development of cancer or damage to children born of people who have eaten irradiated foods.

Critics have also questioned the effects of irradiation on the physical and nutritional value of foods. They suggest that foods treated with radiation may develop unpleasant tastes and textures and that they may not survive on the grocer's shelf as long as foods preserved by other methods. They also wonder whether vitamins, minerals, and other necessary nutrients are destroyed by radiation.

Scientific Evidence and Experience with Food Irradiation

Food irradiation is not a new technology. The practice began in Europe in the 1920s. It has taken much longer to catch on in the United States. Part of the delay was due to a decision by the U.S. Congress in 1958 that irradiating is equivalent to adding a chemical to food. Under this definition, the U.S. Food and Drug Administration (FDA) is required to make a separate decision each time irradiation is proposed for use with a specific food.

The first such decision came in 1963 when the FDA approved irradiation for the treatment of wheat and flour to kill insects. Since that time, the agency has approved irradiation for eight other applications, including its use on white potatoes, spices, pork, fresh fruit, and poultry.

On an international scale, food irradiation is a relatively common method for protecting food. It has been approved for general use in 40 nations, although it is commonly practiced in only 28.

The effects of food irradiation have now been studied scientifically in detail. Those studies show little reason for concerns of the type expressed above. Irradiation appears to have less effect on food properties than do most other methods of food processing, such as freezing, dehydrating, or boiling. With regard to taste and texture, tests have shown that consumers are unable to distinguish between irradiated and nonirradiated foods.

Finally, irradiation is very effective in extending the shelf-life of foods. As an example, nontreated pork can be stored in a refrigerator for a period of about 40 days. By comparison, irradiated pork can be stored under the same conditions for up to 90 days.

The bottom line for many health authorities is the health benefits provided by food irradiation. Even if it turns out that irradiation causes some as-yet-undetected changes in foods, its effectiveness in killing disease-causing microorganisms far outweighs such potential risks. By the end of the 1990s, an impressive list of health-related organizations had expressed support for the use of radiation in preserving foods. Among the organizations are the World Health Organization, the American Medical Association, the American Council on Science and Health, and the American Dietetic Association.

References
Blumenthal, Dale, "Food Irradiation: Toxic to Bacteria, Safe for Humans," *FDA Consumer*, November 1990.

Braus, Patricia, "Food Irradiation" in *The Gale Encyclopedia of Science*, Detroit: Gale Research, 1996, 1515–17.

"Facts about Food Irradiation" <http://www.iaea.or.at/worldatom/inforesource/other/food/index.html> accessed 27 May 1998.

"Food Irradiation Bibliographic Database" <http://www.nal.usda.gov/fnic/foodirad/intro.html> accessed 30 April 1999.

Loaharanus, Paisan, "Cost-Benefit Aspects of Food Irradiation," *Food Technology*, January 1994: 104–08.

Manning, Anita, "Food Safety Regulators Warm to Radiation Idea," *USA Today*, 4 November 1997: 4D.

O'Neil, Patrick, "Digesting Irradiation," *The Sunday Oregonian*, 19 October 1997: E1.

"Position of the American Dietetic Association: Food Irradiation" <http://www.eatright.org/airradi.html> accessed 30 April 1999.

CHAPTER SIX
Documents

The documents in this chapter relate to topics discussed in Chapter 5 on the social issues that have arisen as a consequence of developments in physics research. Some of these documents put forth arguments on one side or the other of certain issues. Other documents present the official positions of governmental bodies or other authorities.

HEALTH EFFECTS OF ELECTROMAGNETIC FIELDS

In 1991, the U.S. Congress asked the National Academy of Sciences to review the scientific literature to determine what risk, if any, low-frequency electrical and magnetic fields posed to human health. The research arm of the Academy—the National Research Council—appointed a group of scholars to serve as the Committee on the Possible Effects of Electromagnetic Fields on Biologic Systems. This committee reported its findings in 1996. The following section is taken from the executive summary of that report.

Conclusions of the Committee

Based on a comprehensive evaluation of published studies relating to the effects of power-frequency electric and magnetic fields on cells, tissues, and organisms (including humans), the conclusion of the committee is that the current body of evidence does not show that exposure to these fields presents a human-health hazard. Specifically, no conclusive and consistent evidence shows that exposures to residential electric and magnetic fields produce cancer, adverse neurobehavioral effects, or reproductive and developmental effects.

The committee reviewed residential exposure levels to electric and magnetic fields, evaluated the available epidemiologic studies, and examined laboratory investigations that used cells, isolated tissues, and animals. At exposure levels well above those normally encountered in residences, electric and magnetic fields can produce biologic effects (promotion of bone healing is an example), but these effects do not provide a consistent picture of a relationship between the biologic effects of these fields and health hazards. An association between residential wiring configurations (called wire codes, defined below) and childhood leukemia persists in multiple studies, although the causative factor responsible for that statistical association has not been identified. No evidence links contemporary measurements of magnetic-field levels to childhood leukemia.

Study Findings

Epidemiology

Epidemiologic studies are aimed at establishing whether an association can be documented between exposure to a putative disease-causing agent and disease occurrence in humans. The driving force for continuing the study of the biologic effects of electric and magnetic fields has been the persistent epidemiologic reports of an association between a hypothetical estimate of electric- and magnetic-field exposure called the wire-code classification and the incidence of childhood leukemia. These studies found the highest wire-code category is associated with a rate of childhood leukemia (a rare disease) that is about 1.5 times the expected rate.

A particular methodologic detail in these studies must be appreciated to understand the results. Measuring residential fields for a large number of homes over historical periods of interest is logistically difficult, time consuming, and expensive, so epidemiologists have classified homes according to the wire code (unrelated to building codes) to estimate past exposures. The wire code classification concerns only outdoor factors related to the distribution of electric power to residences, such as the distance of a home from a power line and the size of the wires close to the home. This method was originally designed to categorize homes according to the magnitude of the magnetic field expected to be inside the home. Magnetic fields from external wiring, however, often constitute only a fraction of the field inside the home. Various investigators have used from two (high and low) to five categories of wire-code classifications. The following conclusions were reached on the basis of an examination of the epidemiologic findings:

- Living in homes classified as being in the high wire-code category is associated with about a 1.5-fold excess of childhood leukemia, a rare disease.
- Magnetic fields measured in the home after diagnosis of disease in a resident have not been found to be associated with an excess incidence of childhood leukemia or other cancers. The link between wire-code rating and childhood leukemia is statistically significant (unlikely to have arisen from chance) and is robust in the sense that eliminating any single study from the group does not alter the conclusion that the association exists. How is acceptance of the link between wire-code rating and leukemia consistent with the overall conclusion that residential electric and magnetic fields have not been shown to be hazardous? One reason is that wire-code ratings correlate with many factors—such as age of home, housing density, and neighborhood traffic density—but the wire-code ratings exhibit a rather weak association with measured residential magnetic fields. More important, no association between the incidence of childhood leukemia and magnetic-field exposure has been found in epidemiologic studies that estimated exposure by measuring present-day average magnetic fields.
- Studies have not identified the factors that explain the association between wire codes and childhood leukemia. Because few risk factors for childhood leukemia are known, formulating hypotheses for a link between wire codes and disease is very difficult. Although various factors are known to correlate with wire-code ratings, none stands out as a likely causative factor. It would be desirable for future research to identify the source of the association between wire codes and childhood leukemia, even if the source has nothing to do with magnetic fields.
- In the aggregate, epidemiologic evidence does not support possible associations of magnetic fields with adult cancers, pregnancy outcome, neurobehavioral disorders, and childhood cancers other than leukemia.

The preceding discussion has focused on the possible link between magnetic-field exposure and childhood leukemia because the epidemiologic evidence is strongest in this instance; nevertheless, many epidemiologists regard such a small increment in incidence as inherently unreliable. Although some studies have presented evidence of an association between magnetic field exposure and various other types of cancer, neurobehavioral disorders, and adverse effects on reproductive function, the results have been inconsistent and contradictory and do not constitute reliable evidence of an association.

Exposure Assessment

[This section discusses the problems involved in measuring the exposure to electric and magnetic fields that people experience.]

In Vitro Studies on Exposure to Electric and Magnetic Fields

The purpose of studies of in vitro systems is to detect effects of electric or magnetic fields on individual cells or isolated tissues that might be related to health hazards. The conclusions reached after evaluation of published in vitro studies of biologic responses to electric- and magnetic-field exposures are the following:

- Magnetic-field exposures at 50–60 Hz delivered at field strengths similar to those measured for typical residential exposure (0.1–10 mG) do not produce any significant in vitro effects that have been replicated in independent studies.

When effects of an agent are not evident at low exposure levels, as has been the case for exposure to magnetic fields, a standard procedure is to examine the consequences of using higher exposures. A mechanism that relates clearly to a potential health hazard might be discovered in this way.

- Reproducible changes have been observed in the expression of specific features in the cellular signal-transduction pathways for magnetic-field exposures on the order of 100 microtesla and higher.

Signal-transduction systems are used by all cells to sense and respond to features of their environments; for example, signal-transduction systems can be activated by the presence of various chemicals, hormones, and growth factors. Changes in signal transduction are very common in many experimental manipulations and are not indicative per se of an adverse effect. Notable in the experiments using high magnetic-field strengths is the lack of other effects, such as damage to the cell's genetic material. With even higher field strengths than those, a variety of effects are seen in cells.

- At field strengths greater than 50 microtesla (0.5 G), credible positive results are reported for induced changes in intracellular calcium concentrations and for more general changes in gene expression and in components of signal transduction. No reproducible genotoxicity is observed, however, at any field strength. Again, effects of the sort seen are typical of many experimental manipulations and do not indicate per se a hazard. Effects are observed in very high field strength exposures (e.g., in the therapeutic use of electromagnetic fields in bone healing).

The overall conclusion, based on the evaluation of these studies, is that exposures to electric and magnetic fields at 50–60 Hz induce changes in cultured cells only at field strengths that exceed typical residential field strengths by factors of 1,000 to 100,000.

In Vivo Studies on Exposure to Electric and Magnetic Fields

Studies of in vivo systems aim to determine the biologic effects of power-frequency electric and magnetic fields on whole animals. Studies of individual cells, described above, are extremely powerful for elucidating bio-chemical mechanisms but are less well suited for discovering complicated effects that could be related to human health. For such extrapolation, animal experiments are more likely to reveal a subtle effect that might be relevant to human health. The obvious experiment is to expose animals, say mice, to high levels of electric or magnetic fields to observe whether they develop cancer or some other disease. The experiments of this sort that have been done have demonstrated no adverse health outcomes. Such experiments by themselves are inadequate, however, to discount the possibility of adverse effects from electric and magnetic fields, because the animals might not exhibit the same response and sensitivities as humans to the details of the exposure. For that reason, a number of animal experiments have been carried out to examine a large variety of possible effects of exposure. On the basis of an evaluation of the published studies in this area, the committee concludes the following:

- There is no convincing evidence that exposure to 60-Hz electric and magnetic fields causes cancer in animals.

There is no evidence of any adverse effects on reproduction or develop-ment in animals, particularly mammals, from exposure to power-frequency 50- or 60-Hz electric and magnetic fields.

- There is convincing evidence of behavioral responses to electric and magnetic fields that are considerably larger than those encountered in the residential environment; however, adverse neurobehavioral effects of even strong fields have not been demonstrated.

Laboratory evidence clearly shows that animals can detect and respond behaviorally to external electric fields on the order of 5 kV/mrems or larger. Evidence for animal behavioral response to time-varying magnetic fields, even up to 3 microtesla, is much more tenuous. In either case, general adverse behavioral effects have not been demonstrated.

- Neuroendocrine changes associated with magnetic-field exposure have been reported;however, alterations in neuroendocrine function by magnetic-field exposures have not been shown to cause adverse health effects.

The majority of investigations of magnetic-field effects on pineal-gland function suggests that magnetic fields might inhibit nighttime pineal and blood melatonin concentrations; in those studies, the effective field strengths varied from 10 microtesla (0.1 G) to 5.2 microtesla (0.052 G). The experimental data do not compellingly support an effect of sinusoidal electric field on melatonin production. Other than the observed changes in pineal function, an effect of electric and magnetic fields on other neuroendocrine or endocrine functions has not been clearly shown in the relatively small number of experimental studies reported.

Despite the observed reduction in pineal and blood melatonin concentrations in some animals as a consequence of magnetic-field exposure, studies of humans provide no conclusive evidence to date that human melatonin concentrations respond similarly. In animals with observed melatonin changes, adverse health effects have not been shown to be associated with electric- or magnetic-field-related depression in melatonin.

- There is convincing evidence that low-frequency pulsed magnetic fields greater than 5 G are associated with bone-healing responses in animals.

Although replicable effects have been clearly demonstrated in the bone-healing response of animals exposed locally to magnetic fields, the committee did not evaluate the efficacy of this treatment in clinical situations.

Source: Reprinted from *Possible Health Effects of Exposure to Residential Electric and Magnetic Fields*, by the Committee on the Possible Effects of Electromagnetic Fields on Biologic Systems, National Research Council, Copyright 1997 by the National Academy of Sciences. Courtesy of the National Academy Press, Washington, D.C.

THE INTERNATIONAL SPACE STATION

For a number of years, space scientists have been thinking about and planning for the next stage in human exploration of space. Over time, a consensus has developed that this next step should involve the construction of a space station orbiting the Earth. Experts view the space station as an opportunity for carrying out experiments in a microgravity environment and, on a long-term basis, as a staging module for future trips to Mars and beyond.

Not all scientists and politicians view the space station as a good idea, however. They raise many of the same issues often argued when very expensive basic research projects are proposed. The two selections below outline the hopes for the space station and the objections raised against it. The first selection below is a general description of the International Space Station prepared by the National Space and Aeronautics Administration (NASA). The second selection is a press release from the office of Representative Tim Roemer (D-IN).

The International Space Station: The Mission

The International Space Station is the largest and most complex international scientific project in history. And when it is complete just after the turn of the century, the station will represent a move of unprecedented scale off the home planet. Led by the United States, the International Space Station draws upon the scientific and technological resources of 16 nations: Canada, Japan, Russia, 11 nations of the European Space Agency and Brazil.

More than four times as large as the Russian Mir space station, the completed International Space Station will have a mass of about 1,040,000 pounds. It will measure 356 feet across and 290 feet long, with almost an acre of solar panels to provide electrical power to six state-of-the-art laboratories.

The station will be in an orbit with an altitude of 250 statute miles with an inclination of 51.6 degrees. This orbit allows the station to be reached by the launch vehicles of all the international partners to provide a robust capability for the delivery of crews and supplies. The orbit also provides excellent Earth observations with coverage of 85 percent of the globe and over flight of 95 percent of the population. By the end of this year, about 500,000 pounds of station components will have been built at factories around the world.

U.S. Role and Contributions
The United States has the responsibility for developing and ultimately operating major elements and systems aboard the station. The U.S. elements include three connecting modules, or nodes; a laboratory module; truss segments; four solar arrays; a habitation module; three mating adapters; a cupola; an unpressurized logistics carrier and a centrifuge module. The various systems being developed by the U.S. include thermal control; life support; guidance, navigation and control; data handling; power systems; communications and tracking; ground operations facilities and launch-site processing facilities.

International Contributions
The international partners, Canada, Japan, the European Space Agency, and Russia, will contribute the following key elements to the International Space Station:

- Canada is providing a 55-foot-long robotic arm to be used for assembly and maintenance tasks on the Space Station.
- The European Space Agency is building a pressurized laboratory to be launched on the Space Shuttle and logistics transport vehicles to be launched on the Ariane 5 launch vehicle.
- Japan is building a laboratory with an attached exposed exterior platform for experiments as well as logistics transport vehicles.
- Russia is providing two research modules; an early living quarters called the Service Module with its own life support and habitation systems; a science power platform of solar arrays that can supply about 20 kilowatts of electrical power; logistics transport vehicles; and Soyuz spacecraft for crew return and transfer.

In addition, Brazil and Italy are contributing some equipment to the station through agreements with the United States.

ISS Phase One: The Shuttle-Mir Program
The first phase of the International Space Station, the Shuttle-Mir Program, began in 1995 and involved more than two years of continuous stays by astronauts aboard the Russian Mir Space Station and nine Shuttle-Mir docking missions. Knowledge was gained in technology, international space operations and scientific research.

Seven U.S. astronauts spent a cumulative total of 32 months aboard Mir with 28 months of continuous occupancy since March 1996. By contrast, it took the U.S. Space Shuttle fleet more than a dozen years and 60 flights to achieve an accumulated one year in orbit. Many of the research programs planned for the International Space Station benefit from longer stay times in space. The U.S. science program aboard the Mir was a pathfinder for more ambitious experiments planned for the new station.

For less than two percent of the total cost of the International Space Station program, NASA gained knowledge and experience through Shuttle-Mir that could not be achieved any other way. That included valuable experience in international crew training activities; the operation of an international space program; and the challenges of long duration spaceflight for astronauts and ground controllers. Dealing with the real-time challenges experienced during Shuttle-Mir missions also has resulted in an unprecedented cooperation and trust between the U.S. and Russian space programs, and that cooperation and trust has enhanced the development of the International Space Station.

Research on the International Space Station
The International Space Station will establish an unprecedented state-of-the-art laboratory complex in orbit, more than four times the size and with almost 60 times the electrical power for experiments—critical for research capability—of Russia's Mir. Research in the station's six laboratories will lead to discoveries in medicine, materials and fundamental science that will benefit people all over the world. Through its research and technology, the station also will serve as an indispensable step in preparation for future human space exploration.

Examples of the types of U.S. research that will be performed aboard the station include:

- Protein crystal studies: More pure protein crystals may be grown in space than on Earth. Analysis of these crystals helps scientists better understand the nature of proteins, enzymes and viruses, perhaps leading to the development of new drugs and a better understanding of the fundamental building blocks of life. Similar experiments have been conducted on the Space Shuttle, although they are limited by the short duration of Shuttle flights. This type of research could lead to the study of possible treatments for cancer, diabetes, emphysema and immune system disorders, among other research.
- Tissue culture: Living cells can be grown in a laboratory environment in space where they are not distorted by gravity. NASA already has developed a Bioreactor device that is used on Earth to simulate, for such cultures, the effect of reduced gravity. Still, these devices are

limited by gravity. Growing cultures for long periods aboard the station will further advance this research. Such cultures can be used to test new treatments for cancer without risking harm to patients, among other uses.

- Life in low gravity: The effects of long-term exposure to reduced gravity on humans—weakening muscles; changes in how the heart, arteries and veins work; and the loss of bone density, among others—will be studied aboard the station. Studies of these effects may lead to a better understanding of the body's systems and similar ailments on Earth. A thorough understanding of such effects and possible methods of counteracting them is needed to prepare for future long-term human exploration of the solar system. In addition, studies of the gravitational effects on plants, animals and the function of living cells will be conducted aboard the station. A centrifuge, located in the Centrifuge Accommodation Module, will use centrifugal force to generate simulated gravity ranging from almost zero to twice that of Earth. This facility will imitate Earth's gravity for comparison purposes; eliminate variables in experiments; and simulate the gravity on the Moon or Mars for experiments that can provide information useful for future space travels.
- Flames, fluids and metal in space: Fluids, flames, molten metal and other materials will be the subject of basic research on the station. Even flames burn differently without gravity. Reduced gravity reduces convection currents, the currents that cause warm air or fluid to rise and cool air or fluid to sink on Earth. This absence of convection alters the flame shape in orbit and allows studies of the combustion process that are impossible on Earth, a research field called Combustion Science. The absence of convection allows molten metals or other materials to be mixed more thoroughly in orbit than on Earth. Scientists plan to study this field, called Materials Science, to create better metal alloys and more perfect materials for applications such as computer chips. The study of all of these areas may lead to developments that can enhance many industries on Earth.
- The nature of space: Some experiments aboard the station will take place on the exterior of the station modules. Such exterior experiments can study the space environment and how long-term exposure to space, the vacuum and the debris, affects materials. This research can provide future spacecraft designers and scientists a better understanding of the nature of space and enhance spacecraft design. Some experiments will study the basic forces of nature, a field called Fundamental Physics, where experiments take advantage of weightlessness to study forces that are weak and difficult to study when subject to gravity on Earth. Experiments in this field may help explain how the universe developed. Investigations that use lasers to cool atoms to near absolute zero may help us understand gravity itself. In addition to investigating basic questions about nature, this research

could lead to down-to-Earth developments that may include clocks a thousand times more accurate than today's atomic clocks; better weather forecasting; and stronger materials.

• Watching the Earth: Observations of the Earth from orbit help the study of large-scale, long-term changes in the environment. Studies in this field can increase understanding of the forests, oceans and mountains. The effects of volcanoes, ancient meteorite impacts, hurricanes and typhoons can be studied. In addition, changes to the Earth that are caused by the human race can be observed. The effects of air pollution, such as smog over cities; of deforestation, the cutting and burning of forests; and of water pollution, such as oil spills, are visible from space and can be captured in images that provide a global perspective unavailable from the ground.

• Commercialization: As part of the Commercialization of space research on the station, industries will participate in research by conducting experiments and studies aimed at developing new products and services. The results may benefit those on Earth not only by providing innovative new products as a result, but also by creating new jobs to make the products.

Assembly in Orbit

By the end of this year, most of the components required for the first seven Space Shuttle missions to assemble the International Space Station will have arrived at the Kennedy Space Center. The first and primary fully Russian contribution to the station, the Service Module, is scheduled to be shipped from Moscow to the Kazakstan launch site in February 1999.

Orbital assembly of the International Space Station will begin a new era of hands-on work in space, involving more spacewalks than ever before and a new generation of space robotics. About 850 clock hours of spacewalks, both U.S. and Russian, will be required over five years to maintain and assemble the station. The Space Shuttle and two types of Russian launch vehicles will launch 45 assembly missions. Of these, 36 will be Space Shuttle flights. In addition, resupply missions and changeouts of Soyuz crew return spacecraft will be launched regularly.

The first crew to live aboard the International Space Station, commanded by U.S. astronaut Bill Shepherd and including Russian cosmonauts Yuri Gidzenko as Soyuz Commander and Sergei Krikalev as Flight Engineer, will be launched in early 2000 on a Russian Soyuz spacecraft. They, along with the crews of the first five assembly missions, are now in training. The timetable and sequence of flights for assembly, beyond the first two, will be further refined at a meeting of all the international partners in December 1998. Assembly is planned to be complete by 2004.

Source: "International Space Station Overview" <http://www.shuttlepresskit.com/ ISS_OVR/index.htm> accessed 30 April 1999.

A Call to Ground the International Space Station

WASHINGTON, D.C.—Comparing the space station to a "giant black hole that is quickly sucking away funding for other promising scientific projects" U.S. Rep. Tim Roemer today renewed his call to end "this budget boondoggle that is rapidly spending billions in taxpayer dollars with little progress to show for it."

Roemer's comments came during a hearing by the House Science Committee to review the latest NASA audit of the space station. The NASA audit, completed by Jay Chabrow, the former CEO of TRW Inc., shows that the station is at least $7 billion over budget—and will cost at least $24 billion to complete, up from NASA's $17.4 billion appraisal earlier this year.

At 37 percent, the station's cost overrun is one of the highest for any government project ever. Despite NASA's lavish spending, the space station is more than three years behind schedule. Each additional year of delay threatens to add billions to the program's overall cost.

"The NASA audit of the space station conclusively demonstrates what the scientific community has long expected: that the space station is beset by budgetary overruns and massive scheduling problems," noted Roemer during the hearing. "The more daylight that shines on this program the darker the outlook for its successful completion."

Outside experts estimate the station's cost at levels much higher than NASA. The General Accounting Office (GAO), for example, believes the real cost of the station is closer to $100 billion rather than $24 billion because NASA fails to include the costs of assembling and operating the station.

The Chabrow audit also indicates that the station's troubles will get worse before they get better. The Russian Space Agency (RSA) is roughly two years behind schedule to deliver key station components. Also, RSA may not be able to deliver 20 Progress rockets needed to launch the station. Replacement rockets will cost roughly $100 million each. Faced with continuing delays, NASA has admitted that it must decide by the end of May whether to drop Russia from the program and ask Congress to approve a significant overhaul of the station—at a dramatic increase to the overall cost.

"When first conceived, space station promoters offered compelling stories about how the project would provide miraculous results: from curing cancer to acting as the first step toward manned exploration of our solar system," continued Roemer. "Ten years later these promises remain closer to fantasy than reality, and the space station has proven more a hindrance than a help toward advancing the scientific missions that helped justify its original existence.

"The scientific mission of the station has been shrinking just as fast as the costs have been escalating. Since its conception in 1984, the station has been redesigned three times. The latest model would accomplish only two of its eight original scientific missions. Furthermore, many of the remaining goals envisioned for the station could be accomplished aboard unmanned satellites or aboard the space shuttle for a fraction of the cost. A return to the Moon would cost far less."

Roemer also noted that the station's rising costs are threatening to crowd out other promising projects within NASA. Last year, NASA shifted $200 million from other programs to pay for station cost overruns. This year, NASA has asked for the authority to shift a further $375 million. "As future cost overruns kick in, the space station's budget literally will become a black hole, sucking away funding for promising projects such as probes within our solar system, earth sciences, and aeronautics," noted Roemer.

"The space station simply provides too little bang for the buck. The scientific rewards do not justify its ever growing cost—both to NASA and the taxpayer. We have lost the $20 billion of the roughly $100 billion it will cost to build and operate the space station. This should not stop us from saving the remaining $80 billion, four times what has been spent so far. This funding would be better spent by helping to pay down the nation's debt, for other scientific projects, or to otherwise invest in our future.

"NASA hopes the space station will recapture the imagination of Americans: to bring back the days when we crowded around our living room TVs to watch Neil Armstrong. But NASA's dream does not mesh with reality. The space station is a boondoggle that threatens to suffocate funding for other worthwhile projects. Last summer, the Mars Pathfinder again induced Americans to gather around their television sets. If the station goes forward, however, future missions similar to the Pathfinder will never leave the drawing board. Congress should terminate the station now before it's too late. The taxpayers deserve better; science deserves better; and our children deserve better."

Source: Press release from the office of U.S. Representative Tim Roemer, Third Congressional District, Indiana, 6 May 1998.

DISPOSAL AND STORAGE OF NUCLEAR WASTES

Experts, politicians, and members of the general public have been arguing about nuclear waste disposal for nearly half a century. Even long after the federal government had apparently made a "final" decision on the issue by authorizing the construction of a waste disposal site within Yucca Mountain in Nevada, the debate goes on. The two documents reprinted below represent two very different views as to the wisdom of using Yucca Mountain as a storage site for nuclear wastes. The first selection is taken from a book published by the University of California Press dealing with a broad range of issues relating to the storage of radioactive wastes. The second selection is taken from a position paper produced by the Nuclear Waste Project of the State of Nevada. An important goal of the project has been to oppose efforts to use Yucca Mountain as a nuclear waste disposal site.

High-Level Radioactive Waste Management in the United States: Background and Status, 1996 (footnotes omitted)

27.1 Introduction

The United States high-level radioactive waste disposal program is investigating a site at Yucca Mountain, Nevada, to determine whether or not it is a suitable location for the development of a deep mined geologic repository. At this time, the United States program is investigating a single site, although in the past, the program involved successive screening and comparison of alternate locations. The United States civilian reactor programs do not reprocess spent fuel; the high-level waste repository will be designed for the emplacement of spent fuel and a limited amount of vitrified high-level wastes from previous reprocessing in the United States. The legislation enabling the United States program also contains provisions for a Monitored Retrievable Storage facility, which could provide temporary storage of spent fuel accepted for disposal, and improve the flexibility of the repository development schedule.

Yucca Mountain is a mountainous ridge located in the southwestern United States in the southern Great Basin, the largest subprovince of the Basin and Range physiographic province of the United States. The Basin and Range province is that area of southwestern North America that is characterized by more or less regularly spaced subparallel mountain ranges and intervening alluvial basins formed by extensional faulting. The regional climate of the southern Great Basin is typically hot and semi-arid. Generally, the geology of the province can be described as a late Precambrian and Paleozoic continental margin assemblage that has been complexly deformed by the late Paleozoic and Mesozoic orogenies. Western portions of the province are broadly overlain by Cenozoic volcanic rocks; the distinctive physiography is largely a product of the most recent phase of extensional deformation. The alluvial basins are characterized by low rainfall, high evapotranspiration, ephemeral streams and closed hydrologic system, evidenced by the absence of drainage external to the basins. Characteristics such as these were important waste isolation considerations in the selection of Yucca Mountain for site characterization.

The repository design concept is a mined excavation at a depth of approximately 300 meters below the crest of Yucca Mountain and at a distance of approximately 300 meters above the regional groundwater table. The site is in silicic volcanic rocks, comprising alternating layers of welded and non-welded volcanic tuffs. The non-welded tuff underlying the proposed repository horizon contain layers that are extensively zeolitized. The strategy for waste isolation relies on both engineered and natural barriers to provide defense in depth. The strategy for long term waste isolation places primary reliance on, and takes advantage of, the natural barriers, which include the aridity of the site, the unsaturated character of the host rock, and the deep regional water table. All indications are that these geologic conditions have been both spatially and temporally stable for many millions of years.

[Intervening sections of this chapter deal with "Legislative Background," "Yucca Mountain Site Waste Isolation Strategy," "Results of Site Characterization Completed to Date," and "Planned Future Work."]

27.6 Conclusions

Following amendment in 1987 of the legislation authorizing characterization of sites for a repository, the United States' high-level waste program focused on Yucca Mountain in the southwestern United States as the single site under consideration. The attributes of Yucca Mountain that made it technically attractive nearly twenty years ago continue to be the technical underpinnings of the strategy for long term waste containment and isolation. Significant progress has been made in the characterization of Yucca Mountain as a potential site for a mined geologic repository. Conditions encountered in the exploratory studies facility tunnel at repository depth are consistent with expectations of such a facility constructed in the unsaturated zone. Total system performance assessments of the long term behavior of a repository at Yucca Mountain continue to mature, and have provided significant guidance in helping define priorities in the test programs and design solutions for the engineered barriers.

The technical strengths of the Yucca Mountain site depend on limited water available to contact the wastes and a corresponding high potential for isolation of the wastes. Today, the United States regulatory approach to long term compliance is uncertain. While the United States Nuclear Regulatory Commission regulations are in place, the United States Environmental Protection Agency standards for disposal safety are remanded. Actions underway to develop a new standard for disposal safety are reopening issues fundamental to the structure of the regulatory approach.

The United States high-level waste program regulations were, in the past, based on a relatively long time frame of regulatory interest, 10,000 years, and assessed compliance against limits on total system releases at an accessible environment, locate five kilometers from the repository. The National Academy of Sciences recommendation that the United States adopt a dose based standard for postclosure compliance for a repository has raised issues relative to the regulatory time frame, dose and risk, the definition of the reference biosphere, human intrusion and the quantitative treatment of natural processes and events. Deliberation of these issues is expected to be intense and time consuming, and fundamental re-evaluation of the United States approach to long term compliance should not be unexpected.

Technically, the Yucca Mountain site remains attractive because of its great potential to isolate wastes. However, there are significant concerns about the ability to bring to closure a regulatory proceeding that could have to deal with what are unprecedented time frames in the context of regulation. The potential for a geologic disposal standard that could introduce a need to rely on dilution in a closed hydrologic basin to meet a dose based standard takes the United States high-level waste program full circle back to the promulgation of the United States Environmental Protection Agency standards for geologic disposal. In those proceedings, a dose based standard was considered to be inappropriate policy that could increase overall population exposures by encouraging disposal methods that would enhance dilution of any radionuclides released.

The extent to which the Yucca Mountain site eventually can be shown to be in compliance with a regulation that is evolving amid questions about the very nature of the regulatory structure that has been the basis for selection of the site, and assessment of its performance for nearly twenty years, is a significant concern. This reassessment is occurring even as the geologic and engineering disciplines are beginning to evolve data sets that are unprecedented in depth, breadth and specificity for evaluation the Yucca Mountain site for its waste isolation potential.

Source: Reprinted with permission from J. Russell Dyer and M. D. Voegele, "High-Level Radioactive Waste Management in the United States: Background and Status: 1996," in *Geological Problems in Radioactive Waste Isolation* (LBNL-38915), edited by Paul A. Witherspoon, Berkeley: Lawrence Berkeley National Laboratory, 1996.

The Battle Over the Proposed Yucca Mountain Nuclear Waste Dump

The controversy over Yucca Mountain centers around the federal government's plans to turn the site into the nation's first high-level nuclear waste repository. If the plan proceeds, 70,000 metric tons of hazardous radioactive materials from U.S. commercial nuclear power plants will be entombed 80 miles northwest of Las Vegas for the next 10,000 years.

Yucca Mountain is the only site being studied. It was picked largely for political reasons, without any alternatives or contingencies. Many Nevadans believe that the political and economic incentives influencing the federal government and other promoters of the project are too great to allow for an objective evaluation process. Project proponents and the nuclear power industry, on the other hand, appear eager to get the site approved despite significant financial, environmental, and health and safety problems. Should Yucca Mountain not work out, the nuclear industry believes it would be set back decades in its goal to build new nuclear power plants.

Numerous studies, both by federal government scientists and independent contractors, suggest that Yucca Mountain is scientifically unsuited for holding the most dangerous nuclear material and keeping it out of the environment for the extraordinarily long time required.

Since a facility like this that must last for 10,000 years has never been built before anywhere in the world, proponents believe that Nevadans should rely on the federal Department of Energy's safety evaluations and predictions that the facility will leak no more than permitted by regulations.

The controversy over Yucca Mountain also involves fundamental issues of a state's right to determine its economic and environmental destiny and to consent or object to federal projects within its borders. Many people in Nevada believe that forcing the project on the State in the face of strong opposition is not only unfair, but it also may lead to Nevada being singled out for other forms of waste disposal and similarly unwanted facilities. For example, DOE wants to convert the Nevada Test Site into "the largest low-level nuclear waste disposal site in the nation."

The State of Nevada has found it necessary to oppose the use of Yucca Mountain as a high-level nuclear waste repository for a variety of reasons, including:

- Much scientific evidence shows that Yucca Mountain is not suited for the disposal of the most dangerous substance in the world—nuclear waste—in that Yucca Mountain is geologically and hydrologically active and complex.
- The entire process of selecting a national nuclear dump site lacked fairness and scientific objectivity and was based on political expediency.
- Large-scale radioactive releases could occur through a variety of possible scenarios caused by volcanoes, earthquakes or hydrothermal activity at Yucca Mountain.
- Deadly radioactive substances could leak out of the dump and create serious long-term health risks to the citizens of Nevada.
- Accidents happen. Nuclear waste transportation could result in accidents harmful to Nevada's citizens and its image as an attractive place to visit, live or locate a business. Similar impacts could affect other states along the nuclear waste transportation routes.

Despite scientific evidence that suggests Yucca Mountain could not meet the requirements for licensing as a nuclear dump, the U.S. Department of Energy is continuing to study Yucca Mountain as the sole site.

The nuclear power industry and DOE are intensifying efforts to get Yucca Mountain approved. The DOE recently attempted to persuade Congress to strip the State of Nevada of its power to enforce federal and state environmental and other regulations.

The nuclear power trade association, the American Nuclear Energy Council, has pledged to spend $30 million on advertising to convince Nevadans that the Yucca Mountain site and nuclear waste are safe. The Council also is funding lobbying efforts in the State to influence unions, business leaders, members of the media, legislators and others to support the Yucca Mountain project.

Part of this effort has involved offers to pay Nevada millions of dollars to give up its right to disapprove of DOE activities and its rights to mitigation and compensation.

Nevada's Governor Bob Miller called this approach "a cynical new strategy to try and buy Nevada's surrender."

DOE and nuclear industry supporters assert that current plans call only for the study of Yucca Mountain to determine if it is, in fact, a suitable repository site. The State of Nevada and others have noted that DOE intends to actually construct a large portion of the repository in the course of its "study" activity. Nevadans are concerned that site study and the investment of more than $6 billion prior to any suitability determination will create an irresistible incentive to go forward with Yucca Mountain regardless of the site's flaws.

[The document next contains a section defining and describing nuclear wastes.]

Is Yucca Mountain Safe?

The principal question asked by scientists as they study Yucca Mountain as the site for the nation's only high-level nuclear waste dump is: Can Yucca Mountain, a block of welded volcanic ash millions of years old, safely isolate for 10,000 years the tons of deadly radioactive waste generated by the nation's commercial nuclear power plants?

Scientific research has shown that, while the site has several potentially favorable characteristics (such as the relatively deep water table, low average annual rainfall, and low population), it has very significant flaws that make it unsuitable for long-term isolation of high-level radioactive materials:

- The site is affected by recent geologic faulting (32 known earthquake faults) and nearby young volcanoes, evidence of a young and active geologic setting.
- The area has a history of earthquakes, including one in 1932 that registered 7.1 on the Richter scale, the same magnitude as the San Francisco earthquake in 1989. A 5.6 magnitude earthquake struck an area less than 12 miles south of Yucca Mountain in June 1992, followed by several hundred aftershocks. Department of Energy scientists had predicted that an earthquake of this size would occur only once in the 10,000-year life of the repository. It occurred less than four years after Yucca Mountain was selected as the only disposal site to be investigated.
- Many prominent scientists believe that groundwater under the mountain has and could rise up again and flood the facility and create a potentially catastrophic situation.
- The proposed Yucca Mountain dump, by regulation, is designed to release radiation. The only question is when. Government scientists believe they can predict it won't leak more than regulations permit for 10,000 years; many other scientists believe that geologic processes cannot be predicted with enough certainty to meet the requirements for licensing such a facility. That length of time—10,000 years—is almost two times longer than the recorded history of mankind.

Other criticisms of the Yucca Mountain dump site include:

- Yucca Mountain was selected on the basis of political considerations in spite of serious geologic flaws.
- Yucca Mountain is situated within a world-class precious metal mining district. Millions of dollars of gold and silver may be located in the area.
- Economic studies reveal that, if built, the repository and its operational activities such as the transport of thousands of shipments of waste along State highways and railroads could negatively affect future investment in Las Vegas, discourage businesses relocating to the area, and cause tourists to not visit Nevada.
- Nearly eight out of 10 Nevadans oppose building a high-level nuclear waste dump at Yucca Mountain.

[The following sections have been omitted because of space limitations. They can be found in the original document.

- Transportation: Despite Safeguards, Risk is Inherent
- Accidents, Mishaps and Dishonesty: The Commercial Nuclear Power Industry and the Department of Energy in Nevada]

The State of Nevada's Position on the Proposed Yucca Mountain Dump

The current high-level nuclear waste dump program is fatally flawed.

- The Nevada site was selected because of political considerations, not on scientific and technical criteria.
- There are no back-up or alternative sites being evaluated along with Yucca Mountain.
- Therefore there are no other sites for comparison.
- The Department of Energy must find Yucca Mountain "suitable" regardless of any technical defects because of pressure from the nuclear power industry and Congress (i.e., there are no alternatives).
- It is wholly unrealistic to expect the Department of Energy to spend $6.5 billion "characterizing" Yucca Mountain and then simply walk away after serious flaws are found.
- Study plans and engineering designs strongly suggest that the Department of Energy's site characterization program is, in fact, geared toward building the first portion of the dump, not merely "studying" the site, as it claims.

The Department of Energy's credibility is so low, especially with respect to waste issues, that it is probably not capable of carrying out a program like the repository.

- Almost two-thirds of the people in Nevada (64%) do not trust the Department of Energy to be honest in reporting results of its Yucca Mountain work, according to polls.
- Nationally, the Department of Energy has major credibility problems due to mismanagement of waste and contamination at almost all defense facilities. This credibility problem is likely to increase as new revelations occur about past and current practices such as the recent General Accounting Office report about continuing cover-ups at the Hanford, Wash., nuclear facility.

Nevada believes that the waste should be stored in dry casks at existing reactor sites for several decades.

- In 1982, when the Nuclear Waste Policy Act was passed, dry storage technology had not been perfected. Now the Nuclear Regulatory Commission accepts it and utilities use it.
- The Nuclear Regulatory Commission has said that it is safe and feasible to store spent fuel at reactor sites for up to 120 years or more.
- Long-term, dry storage can be accomplished using existing resources by allowing half of the fee collected by act of Congress from a tax on electricity generated by nuclear power plants to be used by utilities for on-site storage. The other half could be set aside for a future repository program.
- Such an arrangement would simplify eventual disposal by allowing the waste to cool thermally and radioactively before being shipped, handled, and disposed of.
- The additional time will allow for the development of new technologies which could significantly moderate the entire nuclear waste problem.
- The Department of Energy would have the time needed to rebuild trust and confidence by demonstrating it can clean up existing defense waste sites that are contaminated.
- There would be sufficient time to build public understanding and trust for future facility siting efforts and the eventual identification of a negotiated location for a disposal site.
- This would allow spending at least 40–60 years actually searching for a scientifically sound, publicly acceptable approach to the solution of the nuclear waste problem.

Source: The Nevada Nuclear Waste Project Office, "Why Nevada Is Opposed to Yucca Mountain."

THE COMPREHENSIVE TEST BAN TREATY

Very few individuals enthusiastically support nuclear weapons development. Some people believe, however, that the United States must develop and maintain a substantial nuclear arsenal in order to protect itself against potential enemies and maintain order throughout the world. Others argue that nuclear weapons are so horrible that no good reason can be evinced to support their construction. It is these two general positions that are frequently voiced, along with other arguments pro and con, when international treaties to limit nuclear weapons are proposed. Two views about the need for ratification of the Comprehensive Test Ban Treaty, signed by the United States and many other nations of the world, are expressed below. The first statement is from the Heritage Foundation, a conservative research and educational institute that promotes a strong national defense. The second statement is from Acting Under-Secretary of State John Holum, who advocates banning the testing of nuclear weapons.

Position Statement by the Heritage Foundation

The Clinton Administration is expected to demand that the Senate move quickly to take up the Comprehensive Test Ban Treaty (CTBT), which would prohibit the kind of nuclear weapons testing that India undertook last week. This treaty has profound implications for the security of the United States, and the Senate should not allow itself to be bullied by the Administration. By taking the time to conduct a careful review of the CTBT, the Senate will discharge its responsibility to make a well-informed judgment on this complex and far-reaching treaty, which took four decades to negotiate. The Senate should move deliberately for several reasons:

- Reason #1: India's recent nuclear tests make it clear that the CTBT will not enter into force in the foreseeable future. Ratification by 44 specified states is required for the treaty to enter into force. Three of them—India, North Korea, and Pakistan—have not even signed the treaty, and only six have formally ratified it. India stated at the conclusion of the CTBT negotiations in 1996 that it had no intention of signing the treaty. As a practical matter, India's series of nuclear tests on May 11 and May 13 have put a global ban on nuclear testing out of reach for some time to come.
- Reason #2: By moving carefully, the Senate can better determine whether the CTBT will undermine America's nuclear deterrent. The Senate must determine whether the Department of Energy and the national laboratories can guarantee the safety and reliability of America's nuclear arsenal under test ban strictures. It must therefore take the time to review the Department of Energy's Stockpile Stewardship and Management Program. It also must assess whether the ban on the construction of new nuclear weapons and replacements for America's aging arsenal will endanger national security.
- Reason #3: The Clinton Administration is using a 1999 examination conference to force early consideration. Ratifying states may request

a conference in 1999 to examine the reasons why others have not ratified the CTBT. Administration officials are demanding early approval of the treaty so the United States can participate. By itself, the conference will have little impact. There are fears, however, that the Clinton Administration will use the conference to modify treaty provisions to allow the CTBT to go into effect without holdout states India, North Korea, and Pakistan. This would make a mockery of the Senate's consideration of the CTBT and ignore Senate prerogatives to review amendments to treaties under the advice and consent process.

- Reason #4: The Administration is attempting to implement provisions of the CTBT before it is ratified. The United States is participating in— and funding—the treaty implementation activities of the Vienna-based Preparatory Commission of the Comprehensive Nuclear-Test-Ban Treaty Organization. Further, the U.S. is helping to maintain existing monitoring stations and establish new stations that will be used to verify treaty compliance. Some of these facilities are located on U.S. territory. In each case, the Clinton Administration is acting as if the Senate already had consented to the treaty.

- Reason #5: History demonstrates that the Senate is often justified in giving deliberate consideration to far-reaching arms control agreements. Arms control agreements frequently assume that the existing security environment will remain unchanged for some time to come. Unforeseen events, such as India's nuclear tests, can leave the United States in the position of having to honor arms control agreements that actually undermine its security in a new era. One example is the 1972 Anti-Ballistic Missile (ABM) Treaty, which barred the United States from deploying an effective missile defense system. The United States continues to honor this treaty, even though the other treaty partner (the Soviet Union) no longer exists, and despite the emergence of new threats unforeseen in 1972, such as the risk of accidental or unauthorized missile launches from Russia and the proliferation of missile technology to rogue states such as Iran and North Korea.

- Reason #6: The Senate has more pressing business on its treaty agenda. Several items on the Senate's treaty agenda require its immediate attention. These include a package of three agreements related to the ABM Treaty, signed by a U.S. delegation in New York last September, and the 1997 Kyoto Protocol to the United Nations Framework Convention on Climate Change. The Administration is attempting to bypass the Senate by implementing provisions of these treaties without Senate review. In the case of the agreements related to the ABM Treaty, the Clinton Administration has yet to send them to the Senate for consideration, despite President Clinton's clear commitment to do so. The Senate has no choice but to step in and assert its prerogatives in the face of the constitutionally suspect actions of the Clinton Administration.

India's recent nuclear tests have proven the need for a careful and considered review of the CTBT. The issues surrounding ratification of the treaty are complex, and the implications for America's national security are profound. Members of the Senate should proceed without haste and without apology as they undertake the important job that the Constitution has entrusted to them.

Source: Statement prepared by Baker Spring, Senior Policy Analyst with The Kathryn and Shelby Cullom Davis International Studies Center at The Heritage Foundation. Policy statements, documents, reports, and other material may be ordered from The Heritage Foundation at 214 Massachusetts Ave., NE, Washington, DC 20002-2999, by calling the Publications Office at 800-544-4843, or on the Internet at <http://www.heritage.org/bookstore>.

Senate Testimony of Acting Under-Secretary of State John Holum

The CTBT overwhelming serves our national interest. Let me describe how it does so.

First, by constraining the development of more advanced nuclear weapons by the declared nuclear powers, the CTBT essentially eliminates the possibility of a renewed arms competition such as characterized the Cold War. Without the ability to conduct nuclear explosive tests, none of the weapon states will be able to develop, with high confidence, new, more advanced weapons. For prudent military planners, this means that advanced new types of nuclear weapons will be precluded. . . .

Second, the CTBT also is a nonproliferation treaty. It will erect a further barrier to the development of nuclear weapons by states hostile to our interests and others. Even if a non-nuclear weapon state were able to assemble sufficient nuclear material to produce a simple fission weapon, the CTBT would force it to place confidence in an untested design (which military leaders might find unacceptable), and it would constrain the development of nuclear weapons beyond simple fission designs. Without access to testing data, a would-proliferator cannot develop with any degree of confidence a compact boosted weapon. Design of a two-stage thermonuclear weapon is even more complicated, and confident development even more dependent on test data. . . .

Third, quite apart from the sheer technical obstacles to nuclear weapon development posed by a CTBT, the existence of the Treaty will strengthen international nonproliferation standards and the Nuclear Non-Proliferation Treaty regime, and give the U.S. a stronger hand to lead the global nonproliferation effort. . . .

The fourth reason to ratify the CTBT is that it is effectively verifiable. The U.S. successfully fought for tough verification provisions in the negotiations and would not have signed the Treaty if it were not effectively verifiable. . . .

This brings me to a fifth reason to ratify the Treaty: it will improve our nuclear test monitoring capabilities.

The CTBT augments the current national technical means for monitoring worldwide nuclear testing with additional tools and data not previously available to the United States. It is a net plus. The CTBT establishes global networks of four different types of sensors—seismic, hydroacoustic, radionuclide, and infrasound—that can detect explosions in different physical environments. These networks, comprising 321 monitoring stations, are called the International Monitoring System (IMS). Data will be coming in continuously from IMS. Some of this [sic] data will be recorded at stations in sensitive parts of the world to which we would not otherwise have access. Consider, for example, that the IMS includes 31 monitoring stations in Russian, 11 in China, and 17 in the Middle East. . . .

Sixth, the CTBT will allow us to maintain a safe and reliable nuclear deterrent.

As a condition of U.S. support for a zero-yield CTBT in the summer of 1995, President Clinton announced safeguards which collectively recognize and protect the continued important contribution of nuclear weapons to U.S. national security. The first safeguard mandated the conduct of a Stockpile Stewardship program—for which there must be sustained bipartisan support from Congress—to ensure a high level of confidence in the safety and reliability of our nuclear weapons stockpile. . . .

[With regard to the second safeguard:]

If, in the unlikely event doubts about our ability to maintain the arsenal under a CTBT arise at some point in the future, the Treaty provides for withdrawal from the Treaty if a party decides that its supreme interests are jeopardized. President Clinton has decided (and stated as one of the Safeguards that condition U.S. support for the Treaty) that the safety and reliability of our nuclear weapons is a supreme national interest.

To implement this condition, the President established a certification process requiring the weapon design laboratories and the Department of Defense to review annually all nuclear weapon types. The Secretaries of Energy and Defense, based on the independent advice of the laboratory Directors, the Nuclear Weapons Council, and the Commander-in-Chief of Strategic Command, are required to report annually to the President whether the U.S. nuclear weapons stockpile is, to a high degree of confidence, safe and reliable. If our nuclear deterrent cannot be so certified, the President, in consultation with the Congress, has made it clear that he would be prepared to withdraw from the Treaty under the "supreme interests" clause in order to conduct whatever testing might be required. . . .

At its very core, here is what the CTBT issue comes down to, what the Senate must consider when making its decision: the nuclear arms race is over; arsenals are shrinking; our dramatically fewer remaining weapons can be kept safe and reliable by other means; we don't need tests; proliferators do; the American people overwhelmingly want testing banned.

Source: Hearings before the Senate Committee on Government Affairs, International Security, Proliferation & Federal Services Subcommittee, 18 March 1998. Also at <http://www.acda.gov/ctbtpage/test.htm> accessed 30 April 1999.

IRRADIATION OF FOODS

The two documents that follow present some of the most common arguments in favor of and opposed to the use of radiation for the preservation of foods. The World Health Organization (WHO) is an international organization whose objective is to attain for all people the highest possible level of health. Food & Water is a community environmental organization based in Vermont interested in a variety of safe food issues, including irradiation and pesticide use.

Position Statement by the World Health Organization

Strictly from the scientific point of view, no ceiling should be set for food irradiated with doses greater than the currently recommended upper level of 10 kGy by the Codex Alimentarius Commission (kGy is a unit of measurement for the amount of radiation absorbed by a body). The food irradiation technology itself is safe to such a degree that as long as sensory qualities of food are retained and harmful microorganisms are destroyed, the actual amount of ionizing radiation applied is of secondary consideration. That was the main conclusion of a week-long meeting on high dose irradiation (Geneva, 15–20 September 1997) organized jointly by the World Health Organization (WHO), the United Nations Food and Agriculture Organization (FAO) and the International Atomic Energy Agency (IAEA).

"The knowledge of what can and does occur chemically in high dose irradiated foods which derives from over 50 years of research tells us that one can go as high as 75 kGy, as has already been done in some countries, and the result is the same—food is safe and wholesome and nutritionally adequate," comments Dr. Fritz Kaferstein, Director of the WHO Programme of Food Safety and Food Aid.

The participants reviewed all relevant data related to the toxicological, microbiological, nutritional, radiation chemical, and physical aspects of food exposed to doses greater than 10 kGy and came to the unanimous conclusion that the food is safe for consumption. The focus of the meeting was on considering the generic wholesomeness of foods, appropriately treated and packaged, that are irradiated in the range of 10–100 kGy to eliminate all spoilage and pathogenic microbial contaminants.

"Food irradiation is perhaps the most thoroughly investigated food processing technology. We are quite satisfied with the existing scientific evidence that higher doses of radiation can provide wholesome, nutritious and safe foods," says Dr. Terry Roberts, Chairman of the meeting, former Head of Microbiology at the Institute of Food Research, Reading Laboratory, United Kingdom. "Similar to thermal sterilization of food, the dose should be sufficient to produce a shelf-stable and microbiologically safe product depending on the type of food and specific consumer requirements."

The presence in food of harmful microorganisms such as *Salmonella* species, *Escherichia coli* O157:H7, *Listeria monocytogenes*, or *Yersinia enterocolitica* is a problem of growing concern to public health authorities all over the world. In an attempt to reduce or eliminate the resulting risks, national regulations on food safety are being tightened up in many countries. In the United States of America, for example, the Department of Agriculture has recently issued new regulations for meat and poultry requiring that testing for *Escherichia coli* begin in January 1997 and that raw meat and poultry processed by large firms be virtually free of *Salmonella* beginning in January 1998. For some of these products, food irradiation may well be the best method to ensure the absence of these microorganisms. However, in some instances, an upper limit of 10 kGy for the overall average dose could preclude the effective use of the technology.

In the case of irradiation of spices, this need for a greater average dose has already been recognized in several countries. France permits an average dose of 11 kGy for the irradiation of spices and dry aromatic substances, whereas Argentina and the United States of America permit a maximum dose of 30 kGy for this purpose.

Still higher doses are required for sterilization of food, for instance, for immuno-compromized hospital patients. For this purpose, the Netherlands permits an average dose of 75 kGy which implies that some parts of the food are exposed to doses over 100 kGy. Some other countries, such as the United Kingdom, permit the use of this technology without specifying an irradiation dose limit for this application. South Africa has permitted the marketing of shelf-stable meat products irradiated to an average dose of 45 kGy. For over 20 years, U.S. and Soviet astronauts have been enjoying their irradiated foods, indeed choosing them over other food preservation technologies.

As with other food pasteurization and sterilization technologies involving thermal, mechanical, or photonic energy input, the objective of processing with ionizing radiation is to destroy pathogenic and spoilage microorganisms without compromising safety, nutrition and sensory quality. Common to all these processes are subsequent physical and chemical changes, the extent of which differs significantly among them. In comparison to thermally sterilized foods, the amount of chemical change in a radiation sterilized food is uniform and relatively small.

"For high dose food irradiation, as with other methods of food production, it is important to use raw materials of good quality, to provide adequate packaging, to follow proper processing procedures and to have sound record-keeping, to follow good personal hygiene and sanitation practices, and to handle the processed foods appropriately during distribution," explains Dr. Roberts.

The group came to the following overall conclusions. *Doses greater than 10 kGy*:

- will not lead to changes in the composition of the food that, from a toxicological point of view, would have an adverse effect on human health;
- will greatly reduce potential microbiological risk to the consumer;
- will not lead to nutrient losses to an extent that would have an adverse effect on the nutritional status of individuals or populations.

Therefore, foods treated with doses greater than 10 kGy can be considered safe and nutritionally adequate when produced under established Good Manufacturing Practice.

"Given these reassuring conclusions, the World Health Organization hopes that food irradiation will now become more acceptable as a means for the improvement of food safety which remains one [of] the Organization's priorities," says Dr. Fernando Antezana, WHO Assistant Director-General.

At the present time, some 30 countries are using food irradiation technology for processing a variety of food products.

Source: World Health Organization press release WHO/68, 19 September 1997.

Excerpt from *Meat Monopolies: Dirty Meat and the False Promises of Irradiation*

The recent push for food irradiation fails to acknowledge the technology's inherent dangers, its intricate connections to the nuclear industry, and the FDA's failure to prove safety. Beginning in 1986, the FDA has given the green light to expose nearly our entire food supply to nuclear irradiation. Since then, staunch citizen opposition has kept the technology out of use. But the recent hamburger recall has led both the food and nuclear industries to push hard for beef irradiation's approval. Its use in the beef industry would open the door to irradiation as the "solution" to contamination crises in all food groups, from poultry to fruits and vegetables.

With beef irradiation on the fast-track through the FDA process, citizen opposition, not government regulation, remains the critical component in keeping irradiated food off store shelves. And from the hazards inherent in the technology to the FDA's own admission that the safety studies are flawed, the risks involved with food irradiation still far outweigh the presumed "benefits."

Irradiation Basics

Food is irradiated using radioactive gamma sources, usually cobalt 60 or cesium 137, or high energy electron beams. The gamma rays break up the molecular structure of the food, forming positively and negatively charged particles called free radicals. The free radicals react with the food to create new chemical substances called "radiolytic products." Those unique to the irradiation process are known as "unique radiolytic products" (URPs).

Some radiolytic products, such as formaldehyde, benzene, formic acid, and quinones are harmful to human health. Benzene, for example, is a known carcinogen.

In one experiment, seven times more benzene was found in cooked, irradiated beef than in cooked, nonirradiated beef. Some URPs are completely new chemicals that have not even been identified, let alone tested for toxicity.

In addition, irradiation destroys essential vitamins and minerals, including vitamin A, thiamine, B_2, B_3, B_6, B_{12}, folic acid, C, E, and K; amino acid and essential polyunsaturated fatty acid content may also be affected. A 20 to 80 percent loss of any of these is not uncommon.

Safety Studies Flawed

The FDA reviewed 441 toxicity studies to determine the safety of irradiated foods. Dr. Marcia van Gemert, the team leader in charge of new food additives at the FDA and the chairperson of the committee in charge of investigating the studies, testified that all 441 studies were flawed.

The government considers irradiation a food additive. In testing food additives for toxicity, laboratory animals are fed high levels (in comparison to a human diet) of potential toxins. The results must then be applied to humans with theoretical models. It is questionable whether the studies the FDA used to approve food irradiation followed this process. In fact, the FDA claimed only five of the 441 were "properly conducted, fully adequate by 1980 toxicological standards, and able to stand alone in support of safety." With the shaky assurance of just five studies, the FDA approved irradiation for the public food supply.

To make matters worse, the Department of Preventative Medicine and Community Health of the New Jersey Medical School found two of the studies were methodologically flawed. In a third study, animals eating a diet of irradiated food experience weight loss and miscarriage, almost certainly due to irradiation-induced vitamin E dietary deficiency. The remaining two studies investigated the effects of diets of foods irradiated at doses below the FDA-approved general level of 100,000 rads. Thus, they cannot be used to justify food irradiation at the levels approved by the FDA.

Other studies indicate serious health problems associated with eating irradiated food. A compilation of 12 studies carried out by Raltech Scientific Services, Inc. under contract with the U.S. government examined the effect of feeding irradiated chicken to several different animal species. The studies indicated the possibility of chromosome damage, immunotoxicity, greater incidence of kidney disease, cardiac thrombus, and fibroplasia. In reviewing Raltech's findings in 1984, USDA researcher Donald Thayer asserted, "A collective assessment of study results argues against a definitive conclusion that the gamma-irradiated test material was free of toxic properties."

Studies of rats fed irradiated food also indicate possible kidney and testicular tumors. One landmark study in India found four out of five children fed irradiated wheat developed polyploidy, a chromosomal abnormality that is a good indication of future cancer development.

Irradiation proponents often claim that decades of research demonstrate the safety of food irradiation, but the studies they use to prove it are questionable. For instance, their "proof" includes studies completed by Industrial Bio-Test (IBT), a firm convicted in 1983 of conducting fraudulent research for government and industry. As a result of IBT's violations, the government lost about $4 million and six years of animal feeding study data on food irradiation. Some of this discredited work is still used as part of the "scientific" basis for assurances of the safety of food irradiation.

Accidents Happen

Workers in irradiation plants risk exposure to large doses of radiation due to equipment failure, leaks, and the production, transportation, storage, installation, and replacement of radiation sources. The Nuclear Regulatory Commission (NRC) has recorded 54 accidents at 132 irradiation facilities worldwide since 1974. But this number is probably low since the NRC has no information about irradiation facilities in approximately 30 "agreement states" which have the authority to monitor facilities on their own.

New Jersey is home to the highest concentration of irradiation facilities, and virtually every New Jersey plant has a record of environmental contamination, worker overexposure, or regulatory failures. Accidents can be nearly fatal to workers and extremely dangerous to the surrounding communities. For instance:

In 1991, a worker at a Maryland facility suffered critical injuries when exposed to ionizing radiation from an electron-beam accelerator. The victim developed sores and blisters on his feet, face, and scalp, and lost fingers on both hands.

In 1988, Radiation Sterilizers, Inc. (RSI) in Decatur, GA, reported a leak of cesium 137 capsules into the water storage pool, endangering workers and contaminating the facility. Workers then carried the radioactivity into their homes and cars. Cleanup costs exceeded $30 million, and taxpayers footed the bill.

In 1986, the NRC revoked the license of a Radiation Technology, Inc. (RTI) facility in New Jersey for 32 worker-safety violations, including throwing radioactive garbage out with the trash and bypassing a key safety device. As a result of this negligence, one worker received a near lethal dose of radiation.

In 1982, an accident at International Nutronics in Dover, NJ, contaminated the plant and forced its closure. Radiation baths were used to purify gems, chemicals, food and medical supplies.

In 1974, an Isomedix facility in New Jersey flushed radioactive water down toilets and contaminated pipes leading to sewers. In the same year, a worker received a dose of radiation considered lethal for 70 percent of the population. Prompt hospital treatment saved his life.

Not a Silver Bullet

Irradiation poses serious risks, and it still does not ensure safe meat. Although it kills most bacteria, it does not destroy the toxins created in the early stages of contamination. And it also kills beneficial bacteria which produce odors indicating spoilage and naturally control the growth of harmful bacteria.

Irradiation also stimulates aflatoxin production. Aflatoxin occurs naturally in humid areas and tropical countries in fungus spores and on grains and vegetables. The World Health Organization (WHO) considers aflatoxin to be a significant public health risk and a major contributor to liver cancer in the South.

In addition, irradiation will likely have a mutagenic effect on bacteria and viruses that survive exposure. Mutated survivors could be resistant to antibiotics and could evolve into more virulent strains. Mutated bacteria could also become radiation-resistant, rendering the radiation process ineffective for food exposed to radiation-resistant strains.

Radiation-resistant strains of salmonella have already been developed under laboratory conditions, and scientists at Louisiana State University in Baton Rouge have found that one bacteria occurring in spoiled meat and animal feces can survive a radiation dose five times what the FDA will eventually approve for beef. Scientists exposed the bacteria, called *D. radiodurans*, to between 10 and 15 kilograys (kGy) of radiation for several hours—enough radiation to kill a person several thousand times over. The bacteria, which scientists speculate evolved to survive extreme conditions of dehydration, survived the radiation exposure.

The Nuclear Connection

To irradiate beef and poultry in the U.S. on a mass scale, hundreds of irradiation facilities would need to be built. Currently, the radiation source for most irradiators is cobalt 60, supplied by the Canadian company Nordion International, Inc. But the only isotope available in sufficient quantities for large-scale irradiation is cesium 137, which is also one of the deadliest. With a half-life of 30 years, cesium 137 remains dangerous for nearly 600 years.

The U.S. Department of Energy (DOE) initially encouraged food irradiation as part of its Byproduct Utilization Program (BUP) created in the 1970s to promote the commercial use of nuclear byproducts. The DOE claimed nuclear byproducts "have a wide range of applications in food technology, agriculture, energy, public health, medicine, and industrial technology," and wanted to "ensure full realization of the benefits of the peaceful atom."

At the same time, it would transfer the burden of nuclear waste from weapons production to consumers—a fact the DOE admitted to the House Armed Services Committee in 1983: "The utilization of these radioactive materials simply reduces our waste handling problem . . . we get some of these very hot elements like cesium and strontium out of the waste."

Not only would this take care of DOE's waste problem, it would develop the technology to reprocess spent nuclear reactor fuel in order to recover cesium 137. The reprocessing would also enable the DOE to recover plutonium, the main ingredient for nuclear weapons.

After the 1988 irradiator accident in Decatur, Georgia, the DOE stopped actively promoting food irradiation and the use of cesium 137. But the store of cesium 137 is ready and waiting.

Irradiation Today

With the FDA's imminent approval of beef irradiation, the irradiation industry is poised to use it as a springboard for flooding the market with a new wave of food irradiation promotion. But to be successful, irradiation proponents must convince retailers that consumers want the technology. The irradiation industry sees education or "consumer training" as the key to citizen acceptance.

In response, scientists at major land-grant universities, with the full support of the USDA, are developing "educational" materials. Iowa State University (ISU), home of one of two publicly held food irradiation facilities in the U.S., developed a pro-irradiation educational video with a $39,000 grant from the USDA Extension Service. The USDA gave grants to projects designed to influence public acceptance of food technologies, specifically food irradiation.

But citizens don't want irradiated foods. Surveys conducted in 1990 and 1994 by HealthFocus, a marketing consulting firm specializing in consumer health trends, found that over 80 percent of consumers were concerned about food irradiation. A study at ISU found when consumers are given solid arguments both for and against irradiation, acceptance of the technology is substantially lower than if they were only given the pro-irradiation side of the story. An August 1997 CBS News poll found nationwide 73 percent of people oppose it, and 77 percent say they wouldn't eat irradiated food.

Citizen aversion to irradiation is so strong, no major supermarket chain will carry irradiated foods, and all the top poultry companies in the nation have stated they will not adopt the technology. The U.S. government may approve its use, but that doesn't mean citizens will believe it's safe, or that they will buy irradiated food.

Source: Excerpted from the Food & Water report *Meat Monopolies: Dirty Meat and the False Promises of Irradiation,* by Susan Meeker-Lowery and Jennifer Ferrara. For a copy of the complete report, contact Food & Water at 1-800-EAT-SAFE. or by writing to Food & Water, Inc.; RR 1, Box 68D; Walden, Vermont 05873.

CHAPTER SEVEN
Career Information

Students in beginning physics courses may sometimes wonder what it is, exactly, that physicists do. As one learns the basic principles of velocity and acceleration, mass and weight, vectors and scalars, simple harmonic motion, momentum and inertia, and other concepts, it may be difficult to see how this knowledge can be applied in real jobs in the everyday world. Indeed, it is the sensitive and skilled teacher who can reveal to students what the world of work for physicists is like while providing the fundamental store of knowledge and skills needed to enter that world.

BASIC AND APPLIED RESEARCH

The purpose of this chapter is to provide a brief introduction to the kinds of careers available to physicists. The first distinction that can be made about such careers is that they tend to deal primarily with basic or applied research. *Basic research* involves the investigation of physical phenomena primarily for the purpose of learning more about the natural world. Determining the means by which heat is transferred from the surface of the Sun to its corona and searching for the top quark are examples of basic research. It is, of course, perfectly possible that basic research can lead to inventions and discoveries that have applications in everyday life. But such results are not the primary motivation for basic research.

Applied research, by contrast, is undertaken to solve specific practical problems. Developing methods for using sound waves to tenderize meat and finding techniques for constructing nanosize electrical systems are examples of applied research. Applied research often yields new discoveries about the nature of matter and energy and, therefore, has a basic-research component. However, the primary goal of applied research is to solve particular problems.

In fact, it is often difficult to distinguish whether a given piece of research should be classified as "basic" or "applied." Fortunately, such distinctions are generally not necessary. A physicist often has no reason to state specifically which type of research he or she plans to undertake in any given situation.

FIELDS OF PHYSICS

The discipline of physics covers a very wide range of subjects, including the nature of matter and energy and the interaction between the two. The longer one studies physics, the more likely he or she is to specialize in increasingly limited topics. Individuals who earn doctoral degrees in physics, for example, have typically found some very specific aspect of physics in which they are most interested. That choice does not mean that they do not have to be familiar with the whole range of basic principles in physics. It just means that they have decided to focus on some specific part of physics for their work.

As an example, some physicists choose to concentrate on the nature of light and other forms of electromagnetic energy, topics encompassed by the field of optical physics. Others choose to specialize in nuclear energy; materials science; acoustics; fluids; plasma physics; high energy physics; elementary particles; aerodynamics; or chemical, atomic, and molecular physics. Generally speaking, research opportunities in both basic and applied studies are available in most, but not all, of these fields.

Astronomy is often regarded as a subfield of physics. At one time, astronomers were severely restricted in the methods by which they could study the planets, the stars, interstellar space, and other astronomical phenomena. There was no way to go to the Sun or another star to make firsthand observations of the events taking place there.

Observational astronomy is, of course, still a very large and important field of study. In recent years, however, physicists have learned how to reproduce at least some of the phenomena that take place in space in their own laboratories. As a results, experimental astrophysics has become a growth field that appeals to many new physicists and astronomers.

Subcategories of Physics

The growth of physics and other fields of science has led to the rise of many new careers in physics. Some of these new occupations have grown out of the even greater specialization of larger fields of physics. For example, physicists interested in solid state (or condensed matter) physics can now choose to focus on superconductivity, crystallography, semiconductors, or nanotechnology. Each one of these subcategories has become large enough and important enough to support a whole community of physicists, often with their own professional organization, their own national conventions and meetings, and their own professional journals.

Another direction in which the field of physics has grown is toward interdisciplinary occupations. One of the oldest and most popular of these interdisciplinary fields is that of chemical physics. As the name suggests, this field involves the application of physical principles to the study of matter. A related occupation is known as physical chemistry, and the distinction between the two is sometimes difficult to determine. Scientists in both disciplines are interested in learning more about the nature of atoms, molecules, and groups of particles by studying them as physical objects that obey familiar physical laws.

Biophysics, geophysics, and astrophysics are other fields in which investigators have chosen to study phenomena that involve the interaction of two (or even more) larger fields of physics. Biophysicists, for example, are interested in applying the principles of physics to obtain a better understanding of biological phenomena, such as the transport of fluids across cell membranes and the flow of fluids through blood vessels.

Cross-Disciplinary Careers

Many individuals discover that they are not primarily interested in conducting research in any of the fields mentioned above. For example, they might find that they are more interested in working with people than with laboratory equipment or theoretical calculations. Such individuals might choose to pursue a career in physics teaching rather than some field of physics research.

Some physicists decide to leave research after a period of time. They find that they have become more interested in the administration or supervision of research. They may take a job in industry, government, or academia where they spend time making and carrying out policy decisions rather than operating laboratory equipment.

It is no longer unusual for physicists to work in both industry and academia at the same time. As they make important breakthroughs in their study of matter and energy, they may find that these breakthroughs have practical applications that they can take advantage of by creating their own companies. The biographical sketches of many physicists today are likely to list the companies they own and the patents they hold as a result of their research work.

A degree in physics also can be applied in other fields. For example, patent attorneys are lawyers who do research on existing patents and advise clients about the process of patenting their own discoveries. As more and more physicists choose to start their own companies and/or obtain patents on their own discoveries, patent attorneys will be needed who can understand the science involved in patent applications and appeals.

Museums also provide job opportunities for physicists. Someone with a background in physics is often needed to design exhibits, teach classes at the museum, conduct tours, give talks, visit schools, and carry out the museum's role in teaching about physical phenomena in other ways.

Banks and financial institutions also employ men and women with backgrounds in physics. These individuals advise other officers of the bank or financial institution as to the wisdom of investing in various types of research and development projects, in new equipment, and in other products of physical research.

EDUCATIONAL BACKGROUND

Job opportunities in physics are available for individuals with almost every academic background. Those who are interested in a research career typically complete an extensive training that includes bachelor's, master's, and doctoral degrees, followed by at least a year of postdoctoral study with an existing research team. After this program, researchers usually find work in academic institutions, industrial laboratories, or government facilities.

In 1996, the year for which the most recent data are available, slightly less than half of all physicists with a doctoral degree are actually working outside of the field itself. The American Institute of Physics has attributed this phenomenon to "tight markets in physics, exciting challenges in other fields and normal career evolution" (Amy Schaub, ed., *Preparing Physicists for Work*, AIP, College Park, MD, September 1997: 7). For those who are employed in physics-related jobs, about half hold positions in academic institutions, about 15 percent work in industry, about 20 percent work for national laboratories, and about 10 percent are employed by other organizations.

But those who choose to complete their educational training with less than a doctoral degree can find jobs in physics, too. They may find positions assisting in research and development in industry, in non-research jobs with industry or the federal government, or as high school science teachers. Those with a master's degree can expect similar opportunities with somewhat greater responsibility, such as the staffing and direction of college and university physics laboratories.

Many physics-related jobs require no more than a year or two of postsecondary education. One example is electronic equipment repairers, also known as service technicians or field service representatives. Individuals in this field are responsible for the installation, maintenance, and repair of electronic equipment used in homes, offices, factories, hospitals, and other settings. Among the equipment covered by this service are televisions and computers, telephone systems, industrial equipment, and medical devices.

JOB OPPORTUNITIES AND REMUNERATION

Each year, the U.S. Bureau of Labor Statistics (BLS) attempts to predict what job opportunities will be in various occupations. In the 1998–99 edition of its *Occupational Outlook Handbook*, the BLS has predicted that job openings for physicists and astronomers are likely to decline through the year 2006. The Bureau blames cutbacks in research funds by both the federal government and many parts of industry for this trend. It estimates that the vast majority of new jobs available in physics and astronomy will become available as the result of retirement by current members of the profession. The Bureau warns new graduates in physics and astronomy to expect keen competition for positions.

The BLS relies on a number of sources, including the American Institute of Physics, for its estimates of current earnings in physics and astronomy. According to the most recent (1996) data available from these sources, the median salary for those with a doctorate is $65,000. The median salaries for those with bachelor's and master's degrees are $50,000 and $55,000, respectively. The BLS reports the average salary for physicists employed by the federal government in 1997 to be about $71,800 a year, and for astronomers and space scientists, $77,400 a year.

Earnings in other physics-related fields vary widely depending on the specific occupation, the amount of education, and many other factors. As an example, the average secondary school teacher of physics in the

United States earned $37,300 annually in 1996, although that number varied over a large range, depending on a person's educational background, years of service, geographical location, and other factors.

Similarly, it is virtually impossible to generalize about earnings in physics-related fields that require less than a bachelor's degree. In the field of electronic equipment repair, for example, weekly earnings in 1996 varied from $717 for telephone installers and repairers, to $602 for repairers of electronic communication and industrial equipment, to $573 for those who work with data processing equipment.

SOURCES OF INFORMATION

Additional information about careers in physics is available from a number of print and electronic sources. This information is aimed at nearly every possible audience, from high school students who are simply interested in learning more about jobs in physics to working physicists who have decided to change jobs. The resources listed below are predominately for those who are still thinking about jobs in physics or who have just begun educational programs with that objective in mind.

Print Resources

The obvious first place to begin in learning about careers in physics is the *Occupational Outlook Handbook* (OOH), produced by the Bureau of Labor Statistics and published annually by the U.S. Government Printing Office. OOH covers a great variety of physics-related careers, with each entry including sections on Nature of the Work; Working Conditions; Employment; Training, Other Qualifications, and Advancement; Job Outlook; Earnings; Related Occupations; and Sources of Additional Information. OOH can also be accessed on the Internet at <http://www.bls.gov/ocohome.htm>.

Preparing Physicists for Work: A Career Guide (Amy Schaub, editor; 1997), is a publication of the American Institute of Physics (AIP) intended primarily for working physicists and those just finishing their educational training. Although it focuses largely on job hunting, resume writing, and other job-oriented skills, it also provides a nice overview of careers in physics-related fields. The book can be purchased for $14 from AIP at One Physics Ellipse, College Park, MD 20740, by e-mail at <csv@aip.org>, or by telephone at 301-209-3190.

Other print sources on careers in physics include the following:

Fiske, Peter S., *To Boldly Go: A Practical Guide for Scientists,* Washington, DC: American Geophysical Union, 1996.

National Academy of Science, National Academy of Engineering, and Institute of Medicine, *Careers in Science and Engineering,* Washington, DC: National Academy Press, 1996.

Peters, Robert L., *Getting What You Came For: The Smart Student's Guide to Earning a Master's or Ph.D.*, New York: Noon Day Press, 1992.

Peterson's Engineering, Science and Computer Jobs, Princeton, NJ: Peterson's Guides, 1995.

Rosen, Stephen, and Celia Paul, *Career Renewal: Tools for Scientists and Technical Professionals,* San Diego: Academic Press, 1997.

Tobias, Sheila, Daryl E. Chubin, and Kevin Aylesworth, *Rethinking Science as a Career: Perceptions and Realities in the Physics Sciences,* Tucson, AZ: Research Group, 1995.

An excellent pamphlet on careers in astronomy can be obtained from the American Astronomical Society, AAS Education Office, University of Chicago, 5640 Ellis Ave., Rm AAC112, Chicago, IL 60637. The organization's home page is at <http://www.aas.org>.

An invaluable source of current information on physics careers is the journal *Physics Today*, published by AIP. The journal carries occasional articles dealing with career issues, employment statistics, and other job-related information. But the advertisements it carries may be one of the best overviews of the kinds of jobs available to physics graduates today. The journal is $69 a year to the general reader and can be ordered from the American Institute of Physics, 500 Sunnyside Blvd., Woodbury, NY 11797, or by telephone at 800-344-6902.

Electronic Sources

The number of Web sites on the Internet with information about careers in physics is very large and constantly changing. The following Web pages were among the most valuable at the time this book was written.

<http://www.aip.org/aip/careers/careers.html> is the American Institute of Physics Web page for careers. The site provides information on career information and planning, career workshops, job listings and centers, employer services, and many other topics. This site is probably the first place to look for further information about occupations in physics.

<http://www.nextwave.org/> is *Science* magazine's invaluable Web page for information about careers and career opportunities in all fields of science.

<http://yorty.sonoma.edu/people/faculty/tenn/Jobs.html> is of special interest because it has one of the largest and most complete set of links to information about jobs in physics. The webmaster for this page has done an extraordinary job in compiling specific recommendations for nearly every aspect of physics careers one can imagine. For example, links are provided that will take the user to job information in the fields of acoustics, atomic and plasma physics, aviation and aerospace, beam physics, computational fluid dynamics, crystallography, energy, environmental technology, experimental particles physics, geosciences, health physics, history of science, and more.

<http://ww.phy.ilstu.edu/physjobs.html> is the home page of the Illinois State University Department of Physics. It is of special value because of the links provided to other sources for information about career opportunities in physics and related fields.

<http://physics.utoronto.ca/wiphys/altcar.html> provides information on Alternative Careers for Physicists.

<http://physics.weber.edu/career.html> is a careers and employment page hosted by Weber State University.

<http://phys3.harvard.edu/careers/> is home page for Forum on Creating Careers in Physics, hosted by Harvard University.

Many other colleges and universities have Web pages dealing with information about physics careers and/or about physics degree programs in their own institutions. One can search for specific institutions or ask a search engine to look for "physics jobs," "physics careers," "physics occupations," or similar phrases.

CHAPTER EIGHT
Statistics and Data

Field of Study	1987	1988	1989	1990	1991	1992	1993	1994	1995	1996
Doctoral Degrees in Physics, by Field of Study, 1987–96										
Acoustics	17	16	15	21	13	18	27	20	18	19
Chemical & Atomic/Molecular	79	77	74	87	76	85	95	140	110	129
Electron	6	2	4	2	1	n/a	n/a	n/a	n/a	n/a
Particle	159	174	135	163	182	153	170	176	183	175
Fluids	21	17	14	17	14	17	19	12	18	21
Optics	50	65	78	76	85	94	96	104	98	129
Plasma & High Temperature	72	65	61	42	58	65	62	79	46	48
Solid State & Low Temperature	287	252	296	306	372	408	336	388	371	364
General	238	271	269	323	247	297	340	343	355	324
Other	119	125	127	144	155	163	143	167	166	156

Source: National Science Foundation, Division of Science Resources Studies, *Science and Engineering Doctorate Awards: 1996*, Detailed Statistical Tables, NSF 97-329, by Susan T. Hill (Arlington, VA, 1997), Table 1.

Doctoral Degrees in Physics, by Various Categories, 1987–96 (U.S. Citizens and Permanent Residents)

	Total	Men	Women	Black	Hispanic	Asian
1987	1,137	620	62	3	12	34
1988	1,172	621	60	11	13	47
1989	1,161	596	58	5	14	57
1990	1,265	663	62	4	13	58
1991	1,286	663	76	6	18	50
1992	1,403	696	94	6	26	88
1993	1,399	712	94	7	26	103
1994	1,548	907	137	9	29	255
1995	1,479	916	143	9	26	312
1996	1,485	814	134	12	29	195

Source: National Science Foundation, Division of Science Resources Studies, *Science and Engineering Doctorate Awards: 1996*, Detailed Statistical Tables, NSF 97-329, by Susan T. Hill (Arlington, VA, 1997), Table 3.

Physics Degrees Awarded, 1965–95

	Bachelor's			Master's			Doctoral		
Year	Men	Women	Total	Men	Women	Total	Men	Women	Total
1966	4,384	224	4,608	1,869	80	1,949	976	19	995
1967	4,466	267	4,733	2,015	96	2,111	1,216	32	1,248
1968	4,749	296	5,045	1,993	95	2,088	1,313	25	1,338
1969	5,213	322	5,535	2,139	120	2,259	1,317	32	1,349
1970	5,004	329	5,333	2,047	158	2,205	1,507	37	1,544
1971	4,733	343	5,076	2,042	152	2,194	1,577	48	1,625
1972	4,322	323	4,645	1,876	159	2,035	1,467	38	1,505
1973	3,955	313	4,268	1,642	113	1,755	1,408	50	1,458
1974	3,625	337	3,962	1,526	136	1,662	1,155	51	1,206
1975	3,354	362	3,716	1,453	124	1,577	1,111	58	1,169
1976	3,156	388	3,544	1,319	132	1,451	1,043	44	1,087
1977	3,062	358	3,420	1,193	126	1,319	975	55	1,030
1978	2,961	369	3,330	1,171	123	1,294	884	45	929
1979	2,939	399	3,338	1,184	135	1,319	928	65	993
1980	2,963	434	3,397	1,074	118	1,192	808	54	862
1981	3,009	432	3,441	1,179	115	1,294	844	62	906
1982	3,014	461	3,475	1,128	156	1,284	844	68	912
1983	3,317	483	3,800	1,208	162	1,370	869	59	928
1984	3,361	560	3,921	1,341	194	1,535	915	67	982
1985	3,550	581	4,111	1,333	190	1,523	889	91	980
1986	3,578	611	4,189	1,277	224	1,501	978	100	1,078
1987	3,629	695	4,324	1,300	243	1,543	1,030	107	1,137
1988	3,492	611	4,103	1,428	253	1,681	1,058	114	1,172
1989	3,705	642	4,347	1,448	291	1,739	1,060	101	1,161
1990	3,514	679	4,193	1,523	296	1,819	1,135	130	1,264
1991	3,575	670	4,245	1,441	284	1,725	1,144	142	1,286
1992	3,435	672	4,107	1,539	295	1,834	1,236	167	1,403
1993	3,403	677	4,080	1,463	318	1,781	1,230	169	1,399
1994	3,295	710	4,005	1,655	297	1,952	1,373	175	1,548
1995	3,161	675	3,836	1,535	291	1,826	1,297	182	1,479

Source: National Science Foundation, Division of Science Resources Studies, *Science and Engineering Degrees, 1966-95*, NSF 97-335, by Susan T. Hill (Arlington, VA, 1997), Section C: Detailed Statistical Tables, Table 38.

Federal Appropriations in Physics-Related Fields, 1995–99 (in millions of dollars)					
Agency	1995	1996	1997	1998	1999
NIST[1]	763.8	619.6	581.0	672.9	[2]
NOAA	1,953.4	1,858.8	1,910.8	2,002.1	[2]
NTIA	101.4	54.0	51.7	57.6	[2]
DOD					
Basic research	1,227.0	1,132.5	1,079.8	1,042	1,112
Applied research	3,069.9	2,862.0	2,873.1	2,997	3,189
Other basic defense research	391.1	306.3	337.8	326	354.1
ENERGY R&D					
Solar and other renewable	388.1	275.2	266.3	346.8	365.9
Nuclear	203.1	125.6	219.9	243.1	284.0
Fusion	372.6	244.1	232.4	232.0	229.8
Basic energy	747.3	791.7	649.3[3]	668.2	809.1
High energy physics	646.9	667.0	670.1	680.0	696.5
Nuclear physics	334.7	304	315.9	320.9	335.1
NSF	3,228.7	3,220.0	3,270.0	3,429.0	3,671.2
Research	2,245.0	2,314.0	2,432.0	2,545.7	2770.0
Equipment	126.0	70.0	80.0	109.0	90.0
Education	606.0	599.0	619.0	632.5	662.0
NASA	14,376.7	13,903.7	13,709.2	13,648.0	13,665.0
Science, aeronautics, and technology		5,928.9	5,453.1	5,680.0	5,653.9
Space science	2,012.6	2,032.6	1,969.3	1,983.8	2,119.2
Earth sciences	1,340.1	1,289.4	1,366.6	1,367.3	1,413.8
Micro-gravity	483.1	488.5	243.7	214.2	263.5
Space station	2,100.0	2,100.0	2,148.6	2,501.3	2,270.0
Academic Programs	102.2	106.9	120.4	120.0	138.5
EPA	7,240.9	6,528.0	6,750.0	7,361.0	[4]
Science and Technology	—[5]	525.0	552.0	631.0	650.0

1. For acronyms, please see Glossary.
2. Partial year funding only.
3. Figure represents recategorization of certain subgroups.
4. Not available
5. New program in 1996

Source: U.S. federal budget, information accessed from <http://www.tulane.edu/~aau/>, "Budget and Appropriations."

Institutions Conferring Greatest Number of Doctoral Degrees in Physics and Astronomy, 1996

Institution	Number of Degrees
University of Maryland, College Park	47
University of Texas, Austin	46
University of Illinois, Urbana/Champaign	42
Massachusetts Institute of Technology	41
Princeton University	40
University of Wisconsin, Madison	37
University of Rochester	35
Cornell University	34
Stanford University	34
University of Arizona	34
University of California, Berkeley	34
Ohio State University	29
Harvard University	26
SUNY at Stony Brook	26
California Institute of Technology	25
University of Washington	25
University of California, Los Angeles	24
University of Colorado, Boulder	24
University of Michigan, Ann Arbor	24
University of Chicago	23
University of California, Santa Barbara	22
University of Florida	22
University of Minnesota, Twin Cities	22
Purdue University	21
University of Virginia	21
Columbia University	20
Michigan State University	20
University of Massachusetts, Amherst	20
University of California, San Diego	19
Indiana University, Bloomington	18
University of Pennsylvania	18
University of Pittsburgh	18
City University of New York	17
Northwestern University	17
Rutgers State University, New Brunswick	17
Johns Hopkins University	16
Georgia Institute of Technology	15
University of California, Davis	15

Source: National Science Foundation, Division of Science Resources Studies, *Science and Engineering Doctorate Awards: 1996*, Detailed Statistical Tables, NSF 97-329, by Susan T. Hill (Arlington, VA, 1997), Table 6.

Annual Rate of Pay for Physicists, 1996

Annual Salary	Percent Earning This Range
$27,560–$32,759	4%
$32,760–$40,039	6%
$40,040–$50,439	22%
$50,440–$89,959	45%
$89,960–$124,820	19%

Source: Bureau of Labor Statistics <http://stats.bls.gov/oes/national/oes24102.htm>

CHAPTER NINE
Organizations and Associations

INTRODUCTION

There are dozens of organizations and associations that physicists can join. Some of these organizations consist of scientists from many different fields. The American Association for the Advancement of Science, for example, has members from every field of the biological, physical, earth, and space sciences. It also has members from the fields of politics, sociology, and economics, as well as nonscientists who are interested in science and technology.

A second type of organization is one that enrolls physicists from all fields, such as optics, acoustics, and nuclear physics. These societies are interested in broad issues facing the field of physics in general, as well as the more specialized problems of individual disciplines within physics. The American Physical Society is an example of this type of organization.

Finally, there are organizations such as the Society of Rheology whose members are interested in very special subcategories of physics. These societies tend to focus their agendas on the specific topics of interest to those working in a particular field of physics. Membership is usually not of interest to anyone who is not working in that specialized field of physics.

Like science itself, most organizations of physicists are international in scope. The members of the American Physical Society, for example, come from all over the world. The tie that holds these men and women together is not their nationality, but the field of physics in which they are employed.

Many professional associations in physics have a number of common elements. For example, they often have various categories of membership, including one for individuals and one for corporations. Most groups also have special sections for students, usually with reduced membership rates and often with special publications. Nearly all associations have one or more regularly scheduled national and regional meetings. These meetings are critical to the functioning of the association. They provide members with an opportunity for meeting with each other and exchanging information on recent developments in the field.

This function is also carried out by association publications, such as newsletters, journals, magazines, special reports, and books. Associations generally offer specialized courses for members that allow them to keep up to date with advances in the field. Career counseling and job placement have also become important functions in many organizations. Members are often provided an opportunity for announcing their availability, and corporations are able to announce openings for those working in the field of physics.

Most associations also host awards for individuals who have made important contributions to the field of physics in general or to some specific subcategory of physics.

Finally, many physics groups now host their own Web page. The Web page has become the easiest, quickest, and most complete source of information about the association for specialists in the field as well as for the general public. Most such Web sites have pages open to anyone, while other pages are reserved for member use only.

The following list is not exhaustive. There are far too many individual and specialized societies for all to be included in this chapter. However, the more general organizations are listed. In addition, some typical examples of more specialized societies are also included.

GENERAL ORGANIZATIONS

American Association for the Advancement of Science (AAAS)
1200 New York Ave., N.W.
Washington, D.C. 20005
Telephone: 202-326-6400
e-mail: webmaster@aaas.org
URL: http://www.aaas.org/

The American Association for the Advancement of Science is one of the oldest scientific societies in the United States. It was founded in 1848 in Philadelphia. Over the years, it has served as the parent organization from which many specialized scientific societies have sprung, including the American Chemical Society, the American Anthropological Association, and the Botanical Society of America.

As defined in AAAS's constitution, the organization's mission is to further the work of scientists; facilitate cooperation among them; foster scientific freedom and responsibility; improve the effectiveness of science in the promotion of human welfare; advance education in science; and increase the public's understanding and appreciation of the promise of scientific methods in human progress.

Membership in AAAS is open to anyone who wishes to join. Today, more than 143,000 individuals worldwide belong to the association. These individuals include working scientists, science educators, policy makers, and private citizens from every part of the world.

In addition to its individual members, AAAS has established formal relationships with 285 affiliate organizations. These organizations are scientific and engineering societies from throughout the world. Of these affiliates, 238 are scientific or engineering societies, 44 are state or regional academies of science, and three are city academies. Combined membership of the affiliated organizations exceeds 10 million individuals.

Work of the association is carried out largely through three major directorates: Education and Human Resources, International, and Science Policy. The association also sponsors 24 sections focused on areas of special interest to members. These sections range from the physical, biological, and health sciences to the social, economic, and applied sciences.

To many people, AAAS is probably best known for its publication of the weekly journal *Science*. The journal carries reports of recent research, review articles on topics in scientific research, discussions of public policy relating to science and technology, and fascinating editorial and "letters to the editors" columns.

Association for Women in Science (AWIS)
1200 New York Avenue, Suite 650
Washington, D.C. 20005
Telephone: 202-326-8940
Fax: 202-326-8960
e-mail: awis@awis.org
URL: http://www.serve.com/awis

The goal of the Association for Women in Science is to achieve equity and full participation for women in science, engineering, technology, and mathematics. The organization was created in 1971 and currently has over 5,000 members in all fields of the life, physical, and earth sciences; mathematics; and engineering. Members are organized into 76 local chapters.

A major goal of the organization is to interest girls and women in science and in women's issues. An important tool in carrying out that goal is the bimonthly publication *AWIS Magazine*. AWIS also sponsors a number of conferences and publishes a variety of materials dealing with scientific issues and women's special roles and opportunities in those issues.

GENERAL PHYSICAL SOCIETIES

American Institute of Physics
American Center for Physics
One Physics Ellipse
College Park, MD 20740-3844
Telephone: 301-209-3100
Fax: 301-209-0843
e-mail: aipinfo@aip.org
URL: http://www.aip.org

The American Institute of Physics is an umbrella organization consisting of 10 member associations: the Acoustical Society of America, the American Association of Physicists in Medicine, the American Association of Physics Teachers, the American Astronomical Society, the American Crystallographic Association, the American Geophysical Union, the American Physical Society, the American Vacuum Society, the Optical Society of America, and the Society of Rheology.

AIP was formed in 1931 for three primary reasons. First, the financial hardships of the Great Depression called for finding more economical methods for publishing scientific journals, maintaining membership, and carrying out the other administrative functions of scientific societies. The AIP took over responsibility for these functions for five existing physical societies and later added five more.

Second, AIP provided a means for bringing together the theoretical and practical wings of the physics profession. Advances in modern physical theory had threatened to drive these two fields away from each other, but scientific and political considerations created a need for reuniting the two.

Additionally, AIP assumed responsibility for improving public knowledge about, and education in, the field of physics.

AIP today is governed by a 40-member board whose members are chosen by the 10 member societies. Each society is allotted a certain number of votes according to its relative size. In 1993, AIP moved many of its administrative and publication functions to the newly opened American Center for Physics building at the University of Maryland in College Park.

American Physical Society (APS)
One Physics Ellipse
College Park, MD 20740-3844
Telephone: 301-209-0865
Fax: 301-209-0865
e-mail: membership@aps.org
URL: http://www.aps.org

The American Physical Society was formed in 1899 and is now the largest single professional organization concerned with the field of physics in general. APS has more than 40,000 members worldwide and publishes some of the world's most important physics journals, including *Physical Review*, *Physical Review Letters*, and *Reviews of Modern Physics*.

Many of the society's functions are carried out through divisions, topical groups, forums, and sections devoted to topics of special interest. Examples of APS divisions include Astrophysics, Biological Physics, Chemical Physics, Fluid Dynamics, Laser Science, Physics of Beams, and Plasma Physics. Topical groups deal with even more specific topics, such as Gravitation, Instrument and Measurement Science, Precision Measurement and Fundamental Constants, and Shock Compression of Condensed Matter.

Forums are interdisciplinary groups that deal with the interrelationship of physics with other aspects of society. The five APS forums are concerned with Education, History of Physics, Industrial and Applied Physics, International Physics, and Physics and Society. Finally, APS is also divided into seven geographical sections that provide members with closer contact with professional colleagues. The seven sections are designated as Four Corners (Southwest), New England, New York State, Northwest, Ohio, Southeastern, and Texas.

A free one-year membership in APS is available to undergraduate and graduate students in physics. This membership allows students to learn more about the benefits of belonging to APS.

National Society of Black Physicists (NSBP)
Department of Physics
North Carolina A&T State University
Greensboro, NC 27411-1086
Telephone: 336-334-7646
Fax: 336-334-7283
e-mail: vincent@hepmips.physics.uiowa.edu
URL: http://www.nsbp.org

The National Society of Black Physicists was established in 1977 for the purpose of promoting the well-being of African-American physicists within the scientific community and within society at large. Current membership of the organization is about 100 professional physicists and physics students. Membership categories include regular, associate (for non-physicists), student, and corporate. Local chapters of the organization have been established in Atlanta; Los Angeles; Greensboro, North Carolina; Grambling, Louisiana; Jackson, Mississippi; and Nashville, Tennessee.

NSBP sponsors an annual conference that provides an opportunity for members to exchange information about developments in the discipline, as well as to affirm the opportunity for members to network and socialize with each other.

SPECIALIZED PHYSICS SOCIETIES

Acoustical Society of America (ASA)
500 Sunnyside Boulevard
Woodbury, NY 11797-2999
Telephone: 516-576-2360
Fax: 516-576-2377
e-mail: asa@aip.org
URL: http://asa.aip.org

The Acoustical Society of America was founded in 1929 with an original membership of about 450 scientists interested in the field of acoustics. In 1931, ASA became one of the founding members of the American Institute of Physics. Current membership stands at about 7,000 and includes workers in the fields of music, speech and hearing, noise and noise control, animal bioacoustics, structural acoustics and vibration, underwater propagation, oceanography, robotics and computer sciences, architecture, and the more general fields of biology, physics, and engineering.

One of the important functions of the association has been to develop terminology, measurement procedures, and criteria for determining the effects of noise and vibration on structures and animals. Since 1932, the association has published more than 100 acoustical standards through the American National Standards Institute.

The society holds two annual meetings and publishes the monthly *Journal of the Acoustical Society of America*. In addition, it republishes a number of out-of-print texts in acoustics.

American Association of Physicists in Medicine (AAPM)
One Physics Ellipse
College Park, MD 20740-3846
Telephone: 301-209-3350
Fax: 301-209-0862
e-mail: webmaster@aapm.org
URL: http://www.aapm.org

The American Association of Physicists in Medicine primarily includes members from four fields: therapeutic radiological physics, diagnostic radiological physics, medical nuclear physics, and medical health physics. The overall objectives of these four fields is to use X rays, gamma rays, electron beams, neutrons, and other types of radiation to diagnose and/or treat medical problems.

AAPM has five primary objectives: (1) to promote research and development, (2) to disseminate information about medical physics, (3) to contribute to the education of medical physicists, (4) to encourage and assist in the development of medical physicists, and (5) to promote the use of high-quality medical services.

The association publishes a variety of materials dealing with medical physics, including visual aids, monographs, reports, topical reviews, manuals, proceedings, and other miscellaneous publications. These materials are all available through the Springer-Verlag Publishing Company (orders@springer-ny.com).

AAPM was founded in 1958 as an offspring of the Radiological Society of North America. The association now has about 4,500 members. About 300 of these members serve on one or more scientific, professional, educational, or administrative committees. Examples of such committees include Continuing Education, Public Education, Ethics, Legislation and Regulation, Biological Effects, Nuclear Medicine, Ultrasound, and Finance.

American Association of Physics Teachers (AAPT)
One Physics Ellipse
College Park, MD 20740-3845
Telephone: 301-209-3300
Fax: 301-209-0845
e-mail: aapt-memb@aapt.org
URL: http://www.aapt.org

The aim of the American Association of Physics Teachers is to improve the quality of physics education and to promote scientific literacy in the general public. It consists of teachers at the high school, community college, college, university, and research levels. It was organized in 1930. Among the membership benefits offered by the association are subscriptions to three journals (*Physics Today*, *Announcer*, and either *The American Journal of Physics* or *The Physics Teacher*), discounts on AAPT publications and meetings, group insurance, and one situation-wanted advertisement at no charge in *The Physics Teacher*.

American Astronomical Society (AAS)
2000 Florida Avenue, Suite 400
Washington, D.C. 20009
Telephone: 202-328-2010
Fax: 202-234-2560
e-mail: aas@aas.org
URL: http://www.aas.org

The American Astronomical Society was founded in 1899 and is now the primary professional organization in North America for professional astronomers and others interested in astronomy. The society has a variety of membership categories, including full member, associate member, junior member, as well as emeritus member and three kinds of corporate memberships.

Members of AAS may choose to belong to divisions within the society, such as the Solar Physics Division, Division for Planetary Science, the High Energy Astrophysics Division, or the Historical Astronomy Division.

The society sponsors the publication of three major journals, *Bulletin of the American Astronomical Society*, *The Astronomical Journal*, and *The Astrophysical Journal*.

American Crystallographic Association (ACA)
P.O. Box 96, Ellicott Station
Buffalo, NY 14205-0096
Telephone: 716-856-9600, ext. 379

Fax: 716-852-4846
e-mail: aca@hwi.buffalo.edu
URL: http://www.hwi.buffalo.edu/ACA/

Members of ACA are engaged in research on the crystal structure of proteins, drugs, specialty materials, glasses, liquid crystals, minerals, polymers, and other materials. In addition to classical X-ray studies, modern tools of analysis also include synchrotron radiation and neutron and electron diffraction devices.

Most members of ACA belong to at least one special interest group covering topics such as Materials Science, Biological Macromolecules, Small Angle Scattering, Fiber Diffraction, Amorphous Materials, and General Interest. As part of their association benefits, members receive copies of *Physics Today*, published monthly; *Transactions of the ACA*, published annually; the *ACA Newsletter*, published quarterly; and the *IUCr Newsletter*, published three times a year.

American Geophysical Union (AGU)
2000 Florida Avenue, N.W.
Washington, D.C. 20009-1277
Telephone: 202-462-6900
Fax: 202-328-0566
e-mail: cust_ser@kosmos.agu.org
URL: http://earth.agu.org/

The goal of the American Geophysical Union is to advance understanding of the Earth and its environment in space and to make known to the general public the results of its research. The association currently has about 35,000 members, of whom about one-third are from outside the United States. AGU sponsors a number of regional and national meetings each year, the results of which are published in a variety of its publications. These publications include a weekly newspaper, *Eos;* 17 journals in specialized areas; books; translations; databases; and a variety of electronic services.

AGU was established in 1919 by the National Research Council. For more than 50 years, it operated as an affiliate of the National Academy of Sciences. In 1972, the association was incorporated in the District of Columbia and membership was opened to scientists and students anywhere in the world.

American Nuclear Society (ANS)
555 North Kensington Avenue
La Grange Park, IL 60526
Telephone: 708-352-6622

Fax: 708-352-0499
e-mail: nucleus@ans.org
URL: http://www.ans.org

The American Nuclear Society was founded on 11 December 1954 at a meeting held at the National Academy of Sciences in Washington, D.C. The organization was created to provide a means by which professional scientists with an interest in nuclear science and technology could share their interests and problems.

ANS currently has a membership of about 13,000 individuals from business, education, and government. About 1,000 members live in about 40 countries outside the United States. The Society is organized into 61 local sections in the United States and 9 overseas countries, and also has 32 plant branches, 51 student branches, and 115 corporate members.

The society's main objective is to "promote the advancement of engineering and science relating to the atomic nucleus, and of allied sciences and arts." Some of its specific goals are to encourage research; establish scholarships; disseminate information; hold meetings; integrate the variety of scientific and technological disciplines involved in nuclear research; and cooperate with governmental, educational, and business organizations.

Much of the work of ANS is done through about two dozen committees, including Finance, Revenue, Honors and Awards, Membership, Professional Women in ANS, and Public Policy.

The society publishes a variety of monographs, textbooks, handbooks, directories, proceedings, standards, magazines, and journals. The three peer-reviewed ANS journals are *Nuclear Science and Engineering*, *Nuclear Technology*, and *Fusion Technology*.

American Society for Mass Spectrometry (ASMS)
1201 Don Diego Avenue
Santa Fe, NM 87505
Telephone: 505-989-4517
e-mail: asms@asms.org
URL: http://www.asms.org

ASMS was established in 1969 for the purpose of promoting and disseminating knowledge about mass spectrometry and allied topics. Today, more than 3,500 scientists from academic, industrial, and governmental organizations belong to the organization. Member interests include the development of techniques and instrumentation in mass spectrometry, as well as basic research in chemistry, geology, biological sciences, and physics. The society publishes annually *The Proceedings of the ASMS Conference*, containing abstracts of more than 1,000 papers presented at the conference. An introduction to the subject of mass spectrometry is available in a booklet, "What Is Mass Spectrometry," for sale at the price of $3.00 per copy.

American Vacuum Society (AVS)
120 Wall Street, 32nd Floor
New York, NY 10005
Telephone: 212-248-0200
Fax: 212-248-0245
e-mail: angela@vacuum.org
URL: http://www.vacuum.org/

The American Vacuum Society was created in 1953 in New York City. A group of 56 people with different backgrounds in physics met to discuss problems and applications of high vacuum technology. The group decided to hold a symposium the following year. At the 1957 symposium, the participants organized themselves as the American Vacuum Society.

The association has since grown to a membership of 6,000 individuals from around the world. Much of the organization's work is done through its 8 technical divisions, 4 technical groups, and 20 local-area chapters. Some examples of these divisions are the Thin Films, Vacuum Technology, and Nanometer-Scale Science and Technology divisions and the Biomaterial Interfaces and Electrochemistry and Fluid/Solid Interfaces technical groups.

Among the society's activities are short courses at both basic and advanced levels, a Science Educators Day designed to inform science teachers about vacuum physics issues, and an annual symposium held in the fall of each year. AVS publishes the *Journal of Vacuum Science and Technology*.

Optical Society of America (OSA)
2010 Massachusetts Avenue, N.W.
Washington, D.C. 20036-1023
Telephone: 202-416-1430
e-mail: osamem@osa.org
URL: http://w3.osa.org

The Optical Society of America has a membership of more than 11,400 optical scientists, engineers, and technicians in about 50 countries. The society was organized to increase and disseminate knowledge of pure and applied optics, to promote the common interests of optics researchers, and to encourage cooperation among these researchers. The society maintains 25 local sections in the United States, Canada, and Japan, as well as 21 student chapters in 12 states and South Korea, England, Canada, and Belgium.

OSA sponsors a number of journals and other regular publications, including *Journal of the Optical Society of America (A* and *B), Applied Optics, Journal of Lightwave Technology, Optics Letters*, and *Optics and Spectroscopy*. The society also publishes a quarterly newsletter, *Focal Point*, and a monthly news magazine, *Optics and Photonics News*.

Society for Applied Spectroscopy (SAS)
201-B Broadway Street
Frederick, MD 21701-6501
Telephone: 301-694-8122
Fax: 301-694-6860
e-mail: SASOffice@aol.com
URL: http://www.s-a-s.org

SAS was created in 1956 for the purpose of promoting greater exchange of information among groups of spectroscopists in various regions of the United States. The objective of the society is "to advance and disseminate knowledge and information concerning the art and science of spectros-copy and allied sciences." The society publishes *Applied Spectroscopy* and the electronic edition of *Society for Applied Spectroscopy's Newsletter*.

Society of Rheology (SoR)
500 Sunnyside Boulevard
Woodbury, NY 11797-2999
Telephone: 516-576-2403
Fax: 516-576-2223
e-mail: albertco@umche.maine.edu
URL: http://www.umche.maine.edu/sor/

The Society of Rheology was established in 1929. Its members are interested in the study of the deformation and flow of matter. The word *rheology* was coined in the 1920s to represent this field of physics. Today, membership in SoR stands at about 1,500 individuals from academia, government, and industry. Materials in which members are interested include polymers, metals, petroleum products, rubber, paint, printing ink, ceramics and glass, foods, biological materials, floor preparations, and cosmetics.

SoR sponsors two publications, the *Journal of Rheology*, a bimonthly refereed journal of topics in the field, and *Rheology Bulletin*, a biannual newsletter carrying news of interest to members of the society, such as notices of meetings and abstracts of papers.

CHAPTER TEN
Journals and Internet Resources

The resources listed below are of two general types. Some are devoted to a range of science, part of which is physics. *Scientific American, Science,* and *Science News* are examples of such publications. Others focus strictly on the field of physics in general or on some specific aspect of physics. Examples of these publications are *Physics Today* and *Journal of Chemical Physics,* respectively.

Publications dealing specifically with careers and employment in physics are listed in Chapter 7.

American Institute of Physics Journals
The American Institute of Physics publishes a number of journals in specialized fields of physics. The titles are listed below. All AIP journals can be ordered at the following address:

American Institute of Physics (AIP)
One Physics Ellipse
College Park, MD 20740-3843
Telephone: 301-209-3100
Fax: 301-209-0843
e-mail: aipinfo@aip.org

Applied Physics Letters

Chaos

Computers in Physics

Journal of Applied Physics

Journal of Chemical Physics

Journal of Mathematical Physics

Journal of Physical and Chemical Reference Data

Medical Physics

Physics of Fluids

Physics of Plasmas

Physics Today

Review of Scientific Instruments

American Scientist
American Scientist
P.O. Box 13975
Research Triangle Park, NC 27709
Telephone: 919-549-0097 or 800-282-0444
Fax: 919-549-0090
e-mail: subs@amsci.org

American Scientist is a publication of Sigma Xi, the Scientific Research Society. It is published bimonthly and features a variety of articles about science and technology. Many of the contributions are written by scientists and engineers working at the forefront of their fields, providing a valuable overview of the direction and pace of progress in these areas. An interesting feature of the journal is the once-a-year review of books, software, and other products written by children for younger readers.

American Scientist has also published two books of readings collected from earlier articles in the journal. They are *Exploring Evolutionary Biology* and *Exploring Animal Behavior*.

Applied Optics
Optical Society of America
2010 Massachusetts Avenue, N.W.
Washington, D.C. 20036-1023
Telephone: 202-416-1430
e-mail: osamem@osa.org

This journal is sponsored by the Optical Society of America and deals with topics such as optical technology, information processing, and lasers and photonics.

Applied Physics Letters
See **American Institute of Physics Journals.**

Astronomical Journal
Astrophysical Journal
Bulletin of the American Astronomical Society
The three journals listed above are published by the American
Astronomical Society:

American Astronomical Society
2000 Florida Avenue, Suite 400
Washington, DC 20009
Telephone: 202-328-2010
Fax: 202-234-2560
URL: http://www.aas.org

The first two journals carry refereed papers in the field of astronomy
and astrophysics, respectively, while the Bulletin is used primarily for
association news, such as announcements and agendas of meetings,
obituaries, annual reports, and abstracts of meeting sessions.

Chaos
See **American Institute of Physics Journals.**

Computers in Physics
See **American Institute of Physics Journals.**

Discover
Discover
Circulation Department
114 Fifth Avenue
New York, NY 10011-5690
Telephone: 800-829-9132 or 515-247-7569
e-mail: discover@enews.com

Some publishers have long felt that there is a need for a magazine about
science that would appeal to the science-interested general public the way
Sports Illustrated fills the needs of sports fans. A number of such magazines have been initiated, but the Walt Disney Company's *Discover* has
probably been the most successful glossy example. It is published 12 times
a year and carries stories that attempt to make advances in science and
technology understandable to the general public. The magazine claims to
be "the only magazine that covers the entire world of science."

In addition to the magazine itself, *Discover* provides a number of other print and nonprint resources. For example, it maintains a Discover Magazine Virtual Bookstore from which books can be purchased online, and it has published two video series, Great Minds of Science and Secrets of Science. *Discover* also appears as a weekly television program that focuses on a specific topic in science and technology. In one month, for example, topics of the television show were "Air Disasters," "Mutations," "Size and Scale," and "Born to Kill?"

Fusion Technology
American Nuclear Society
555 North Kensington Avenue
La Grange Park, IL 60526
Telephone: 708-579-8287
e-mail: scipubs@ans.org

Fusion Technology is published by the American Nuclear Society and is devoted to reports on theoretical and applied research in the field of nuclear fusion, with special emphasis on its potential use as a source of energy in the future.

The Industrial Physicist
The Industrial Physicist
One Physics Ellipse
College Park, MD 20740-3843
Telephone: 301-209-3100
Fax: 301-209-0843
e-mail: aipinfo@aip.org

The Industrial Physicist is a semi-popular magazine devoted to the development, advancement, and recognition of industrial physicists and the applications of physics in industry.

Journal of Applied Physics
See **American Institute of Physics Journals.**

Journal of Chemical Physics
See **American Institute of Physics Journals.**

Journal of Geophysical Research
American Geophysical Union
2000 Florida Avenue, N.W.
Washington, D.C. 20009-1277
Telephone: 202-462-6900 or 800-966-2481
Fax: 202-328-0566

e-mail: service@agu.org

The *Journal of Geophysical Research* is the largest and best known of 10 peer-reviewed journals sponsored by the American Geophysical Union. It carries a variety of original research reports on the physics and chemistry of the Earth, its environment, and the Solar System.

Journal of Mathematical Physics
See **American Institute of Physics Journals.**

Journal of Physical and Chemical Reference Data
See **American Institute of Physics Journals.**

Journal of Rheology
Journal of Rheology
One Physics Ellipse
College Park, MD 20740-3843
Telephone: 301-209-3100
Fax: 301-209-0843
e-mail: subs@aip.org

The *Journal of Rheology* is the official peer-reviewed publication of the Society of Rheology. It is published six times a year.

The Journal of the Acoustical Society of America
Acoustical Society of America
Circulation and Fulfillment Division
American Institute of Physics
500 Sunnyside Blvd.
Woodbury, NY 11797-2999
Telephone: 516-576-2270 or 800-344-6902
e-mail: subs@aip.org

The Journal of the Acoustical Society of America is the official publication of the Acoustical Society of America. It carries reports on theoretical and experimental research in many fields of acoustics, such as underwater sound and acoustical oceanography, ultrasonics and quantum acoustics, architectural and structural acoustics, bioacoustics, music and noise, and psychology and physiology of hearing.

Journal of the Optical Society of America (A & B)
Optical Society of America
2010 Massachusetts Avenue, N.W.
Washington, D.C. 20036-1023
Telephone: 202-416-1430
e-mail: osamem@osa.org

Volume A of this publication deals with topics related to vision, such as atmospheric optics, scattering and coherence theory, image processing, and machine vision. Volume B deals with optical physics topics, such as fiber optics, modern quantum optics, spectroscopy, and lasers.

Journal of Vacuum Science & Technology (A & B)
Journal of Vacuum Science & Technology
Caller Box 13994
10 Park Plaza, Suite 4A
Research Triangle Park, NC 27709
Telephone: 919-361-2787
Fax: 919-361-1378
e-mail: jvst@jvst.org

The *Journal of Vacuum Science & Technology* is the official publication of the American Vacuum Society. Volume A of the journal is devoted to vacuums, surfaces, and films, while Volume B deals with microelectronics, processing, and phenomena.

Kids Discover
Kids Discover
P.O. Box 54205
Boulder, CO 80322-4205
212-242-5133

Kids Discover is published 10 times a year, with each issue focusing on a single topic. Some topics that have been covered in the past include oil, the Solar System, the brain, garbage, endangered species, and light.

Medical Physics
See **American Institute of Physics Journals.**

Nature
Nature America, Inc.
345 Park Avenue South
10th Floor
New York, NY 10010-1707
Telephone: 800-524-0384
Fax: 615-377-0525
e-mail (subscriptions): subscriptions@nature.com

Nature is the United Kingdom counterpart of *Science* in the United States, that is, a highly respected journal covering all areas of science. It provides news and research reports that are accessible to the general reader with a moderate background in science, along with more esoteric reports on research in all fields of science.

The journal now publishes specialized magazines in the fields of medicine (*Nature Medicine*), genetics (*Nature Genetics*), biology (*Nature Structural Biology*), and biotechnology (*Nature Biotechnology*).

The family of *Nature* journals is also available online: <http://www.nature.com/>. This URL is open to the general public and provides weekly tables of contents, summaries, news articles, access to *Nature* archives, and job information.

New Scientist

New Scientist
1150 185th Street, N.W., #725
Washington, D.C. 10036
Telephone: 888-800-8077
Fax: 200-331-2082
e-mail: newscide@soho.ios.com

New Scientist is published 51 times a year. It was begun in 1956 "for all those men and women who are interested in scientific discovery and in its industrial, commercial and social consequences." In its first two decades, the magazine was a powerful advocate for the analysis of the social, political, economic, and ethical implications of scientific research and technology. Over the last decade, it has tempered its social agenda to some extent, although it is arguably the best single source for a rounded view of the place of science and technology in modern society, along with being an excellent source of information on developments in both areas.

Nuclear Science and Engineering

American Nuclear Society
555 North Kensington Avenue
La Grange Park, IL 60526
Telephone: 708-579-8287
e-mail: scipubs@ans.org

Nuclear Science and Engineering has been published since 1956 and carries information on research on the peaceful use of nuclear energy and radiation.

Nuclear Technology

American Nuclear Society
555 North Kensington Avenue
La Grange Park, IL 60526
Telephone: 708-579-8287
e-mail: scipubs@ans.org

Nuclear Technology is a journal published by the American Nuclear Society devoted to topics such as reactor technology, operations, safety materials, instrumentation, fuel, waste management, medical uses, radiation detection, production of radiation, health physics and computer applications.

Optics Letters

Optical Society of America
2010 Massachusetts Avenue, N.W.
Washington, D.C. 20036-1023
Telephone: 202-416-1430
e-mail: osamem@osa.org

The Optical Society's "quick turn-around" journal covers recent advances in atmospheric optics, quantum electronics, Fourier optics, integrated optics, and fiber optics.

Physical Review

American Physical Society (APS)
One Physics Ellipse
College Park, MD 20740-3844
Telephone: 301-209-0865
Fax: 301-209-0865
e-mail: membership@aps.org

The various editions of *Physical Review* are published by the American Physical Society. The editions are designated by letters and/or numbers to indicate the subject that they cover. For example, *Physical Review A* deals with atomic, molecular, and optical topics; *Physical Review B1*, with condensed matter, structure, phase transitions, non-ordered systems, magnetism, superconductivity, and superfluidity; and *Physical Review D15* with particles, fields, gravitation, and cosmology.

Physics of Fluids
See **American Institute of Physics Journals.**

Physics of Plasmas
See **American Institute of Physics Journals.**

Physics Today

American Institute of Physics
500 Sunnyside Boulevard
Woodbury, NY 11797
Telephone: 800-344-6901 (members) or 800-344-6902 (non-members)
e-mail: aipinfo@aip.org

Physics Today is the most widely distributed general interest journal in physics. It always carries about three major feature articles on topics of special interest in physics as well as a number of regular departments. These departments include Physics Update, Letters, Search and Discovery, Washington Reports, and Physics Community. A few issues each year are devoted entirely to special topics and are larger in size than other issues.

Popular Science
Popular Science
P.O. Box 51282
Boulder, CO 80322-1282
Telephone: 800-289-9399

Popular Science has been published continuously since 1872. It is one of the premier magazines attempting to explain developments in science and, especially, technology to the average reader. Some of the areas covered by the magazine include automotive science and technology, consumer electronics, computers and software, home technology, photography, and aviation and space.

Review of Scientific Instruments
See **American Institute of Physics Journals.**

Science
American Association for the Advancement of Science
1200 New York Avenue, N.W.
Washington, D.C. 20005
Telephone: 202-326-6400
e-mail: webmaster@aaas.org

Science is one of the two premier scientific journals for general readers around the world (*Nature* being the other). The magazine contains seven major sections: News and Comment; Research News; Perspectives; Articles; Reports; Technical Comments; and a Web Feature. In addition, its regular departments include a brief overview of the magazine, This Week in *Science*; an Editorial; Letters to the Editor; brief news notes, Random Samples; Book Reviews; and information on new technology and products, "Tech.Sight: Products."

Science has an extensive Internet Web site that includes two major sections, Science NOW and Science's Next Wave. Science NOW includes Current Stories and Week in Review stories from the world of science, along with a search function that allows a limited search of the journal's archives and current issues. Next Wave includes sections such as Going Public, an open forum for discussion of topics in science; New Niches, a feature on alternative careers in science; In the Loop, the page's news section; Tooling Up, a column on science career advice; Signposts, links to other Web pages; and Wavelengths, an opportunity to interact with editors of *Science*.

Science News

Science News Subscriptions
231 W. Center Street
P.O. Box 1925
Marion, OH 43305
Telephone: 800-552-4412 or 202-785-0931 (collect)
e-mail: scinews@sciserv.org

Science News is a publication of Science Service, a nonprofit corporation based in Washington, D.C. The publication started out in 1921 as a series of news releases dealing with scientific issues. It evolved into the current magazine, published 51 times a year. Science Service also administers the annual Westinghouse Electric Corporation's Science Talent Search and the International Science and Engineering Fair.

The magazine is normally a 16-page publication that summarizes about a dozen important breakthroughs in all fields of science, along with brief Research Notes from two or three major fields of science and two articles covering some major topic. The magazine also includes book reviews and a letters-to-the-editor section. *Science News* is arguably the best single publication for reporting on advances in science available to the general reader.

The Sciences

New York Academy of Sciences
2 East 63rd Street
New York, NY 10021
Telephone: 212-838-0230, ext. 342, or 800-843-6927
Fax: 212-888-2894
e-mail: publications@nyas.org

The Sciences describes itself as "the cultural magazine of science. It combines the literary and aesthetic values of a fine consumer magazine with the authority of a scholarly journal to lead the reader in an enlightening exploration of the world of science." And, indeed, the journal does provide a broad and complex overview of scientific research that is probably not presented as well in any other general-interest magazine.

The appeal of the magazine is reflected in its having won National Magazine Awards in 1986, 1988, 1989, 1991, and 1996, along with a host of other publishing honors.

Scientific American

Scientific American
415 Madison Avenue
New York, NY 10017-1111
Telephone: 800-333-1199 or 515-247-7631
Fax: 515-246-1020
e-mail: customerservice@sciam.com

Scientific American is the oldest and most prestigious scientific journal written for the general reader. It was first published in 1845 and is published monthly. The magazine includes articles covering every field of science and technology, written at a level that an educated and motivated high school or college student or adult can understand.

Scientific American now has an extensive Internet site <http://www.sciam.com/> that includes the current issue, archives of past issues, weekly features, breaking news stories, interviews with scientists, question-and-answer sessions between readers and specialists in all fields of science, and links to other sites containing further information on magazine articles.

Technology Review

Technology Review
P.O. Box 489
Mount Morris, IL 61054
Telephone: 800-877-5320
e-mail: trsubscriptions@mit.edu

Technology Review is published eight times a year by the Association of Alumni and Alumnae of the Massachusetts Institute of Technology. The magazine carries popular accounts of important technological advances, often with special emphasis on social, political, economic, and ethical issues related to the technology. Typical of the stories carried are those dealing with complexity theory, microwave transmission of solar energy, the land mine controversy, and recycling of household garbage.

Some regular departments are a letters-to-the-editor column, the MIT Reporter (An Inside Look at Research at MIT), Trends, The Economic Perspective, and The Humane Engineer.

WonderScience
American Chemical Society
1155 Sixteenth Street, N.W.
Washington, D.C. 20036
Telephone: 800-209-0423
e-mail: r_foster@acs.org

WonderScience is published jointly by the American Institute of Physics and the American Chemical Society. It is a four-color magazine containing creative activities that elementary school teachers can use in the classroom. It features a variety of hands-on activities that can be carried out with inexpensive, safe, and easily accessible materials found at home or in the grocery store. *WonderScience* is published eight times a year, with four issues in September and four in January. A complete set of back issues for the last 10 years of the magazine is also available.

CHAPTER ELEVEN
Glossary and Acronyms

GLOSSARY

The terms included in this chapter are those that may not be familiar to students in their first-year physics course. The reader is also encouraged to refer to any beginning high school or college textbook for more common terms in physics that are not listed here.

"AND/OR" gate   An electronic mechanism that provides a logic 1 ("true") output if and only if two logic 1 ("true") inputs are provided to it. Any inputs other than two "1" inputs results in a logic 0 ("false") output.

antigravity   A hypothesized force in which objects appear to repel rather than attract each other, as in manifestations of gravity.

antimatter   Antimatter is a "mirror image" of normal matter. It has all the same properties as normal matter, except in reversed form. For example, an antiproton is exactly like a proton except that it carries a negative, rather than a positive, charge.

antiparticle   Particles with the same mass that differ from each other in some other fundamental property, such as electric charge or spin.

antiquark   An antiquark is a form of antimatter.

bandgap   The energy difference between two energy levels in a metal.

Big Bang   The term used to describe the event that took place at the birth of the universe when energy was first converted into matter.

Big Crunch   A term used to describe the ultimate collapse of the universe, which is anticipated under certain cosmological theories.

black hole   An astronomical object that is believed to form as the result of the collapse of a supernova and in which gravitational force is so strong that nothing, including radiant energy, is able to escape.

Bose-Einstein condensate   A form of matter that develops at temperatures close to that of absolute zero in which all particles occupy the same quantum state.

chalcogen   A member of Group 16 in the periodic table that includes oxygen, sulfur, selenium, and tellurium.

corona   The outermost region of the Sun's atmosphere.

cosmological constant   A constant invented by Albert Einstein to account for a repulsive force that seemed to exist in the universe.

cosmology   The study of the origin, structure, and fate of the universe.

dark matter   Matter that most astrophysicists believe exists in the universe, although it cannot be observed directly because it is non-luminous.

diquarkonium   A type of exotic meson that consists of two pairs of quarks—two quarks and two antiquarks.

entanglement   A property possessed by two particles such that they are inextricably linked to each other, no matter how far they are separated from each other.

entropy   The tendency of any physical system to become more disordered.

flatness problem   The astrophysical dilemma created by the fact that the universe appears to be flat, but that it may, in fact, exist in other conformations, such as spherical, or in the form of a hyperbolic paraboloid.

fullerene   An allotrope of carbon in which carbon atoms are joined to each other in complex systems of pentagon, hexagon, and/or heptagon rings.

glueball   A type of exotic meson that consists of two or three gluons bound to each other.

gluon   A force-carrying particle that usually exists in a virtual state and is thought to hold subatomic particles together, such as the protons and neutrons in an atomic nucleus.

gravitational lensing   The tendency of massive objects in space to curve light waves that pass near them, thus changing the appearance of an object in much the same way that a glass or plastic lens does.

harmonic conversion   A process that occurs when light from a laser passes through gas in a waveguide, knocking electrons loose from their nuclei. When those electrons return to their ground state in the atom, they release a very large pulse of energy with frequency in the X-ray region.

horizon problem   A dilemma in astrophysics resulting from the observation that all parts of the universe appear to be exactly alike, no matter the direction in which one looks.

inflation theory   A set of astrophysical theories that attempt to describe and explain the changes that took place in the universe in the first few seconds following the Big Bang.

isotropy   The tendency of a transparent material to transmit light with equal intensity in all directions.

lepton   A type of fundamental particle with spin quantum number of 1/2 and with no tendency to experience the strong nuclear force.

logic gate   *See* "AND/OR" gate.

macroscopic scale   The scale of physical phenomena that can be detected with the human senses.

magnetic confinement   A technique for enclosing the heated plasma formed during a fusion reaction by means of very strong magnetic fields.

Massive Compact Halo Object (MACHO)   Objects hypothesized to exist in the halo surrounding our galaxy, with very large mass but little or no luminosity. MACHOs have been thought to be possible candidates for the dark matter that presumably exists in the universe.

meson   A subatomic particle that consists of a quark and an antiquark and that experiences the strong nuclear force.

neutrino   An uncharged elementary particle believed to have little or no mass.

neutron star   An astronomical object believed to form as the final stage in the collapse of a star, with a very large mass and a very large gravitational field.

oscillation   The process in which some physical system changes back and forth between two states periodically. The term has also been used to describe the apparent change in the nature of neutrinos passing through the Earth, in which any given neutrino may exist in any one of three forms at any given moment.

pair production   A physical process in which two particles, a particle and its antiparticle, are formed instantaneously from a burst of energy.

pi bond   A covalent double bond formed by electrons located above and below the plane joining the atoms connected by the bond.

quantum dot   A collection of electrons confined within a semiconductor material, all of which occupy the same quantum state.

quantum mechanics   A quantum mechanical theory that explains the behavior of matter at the submicroscopic level.

quarks   Elementary particles that exist in pairs, such as the top and bottom quarks, and that carry fractional electrical charges in multiples of one-third.

sequestration   The "capturing" of metal ions by means of the formation of coordination compounds, generally by compounds of phosphorus.

serendipity   The unexpected discovery of some phenomenon, often as the result of an accident that occurs during an experiment.

shock wave   A compressional wave of large amplitude formed as the result of some violent physical event, such as an explosion.

smoothness problem   The astrophysical problem arising from the observation that stars, galaxies, and other large concentrations of matter appear to be spread out across the universe in relatively equal concentrations.

solar stalks   Tall, narrow structures located at the top of magnetic loops that originate in the Sun's surface and extend far into the corona.

solar wind   The flow of particles constantly emitted from the Sun's surface that spreads out over much of the Solar System.

space-time   A term used to describe the four-dimensional character of the physical world, consisting of three spatial dimensions (length, width, and height) and one temporal dimension (time).

Standard Model   The physical theory that attempts to explain the fundamental nature of matter by hypothesizing the existence of certain basic particles, such as quarks, leptons, and gluons.

subatomic particle   Any particle smaller than an atom, such as a proton, a quark, or a lepton.

superconductivity   The tendency of a material to lose all resistance to the movement of electrons through it, resulting in the ability of that material to conduct an electrical current essentially forever.

supernova   The explosion of a very large star resulting, according to present theories, in the formation of a black hole.

teleportation   The process by which a particle or group of particles is trans-ferred—essentially instantaneously—from one point to another.

tomography   The production of a three-dimensional image of an object by passing radiation of some type through a body.

Uncertainty Principle   A quantum mechanical principle that it is not possible to determine accurately both the position and the motion of an object.

waveguide   A device by means of which an electromagnetic wave can be confined and directed.

ACRONYMS

AAAS   American Association for the Advancement of Science

AAPM   American Association of Physicists in Medicine

AAPT   American Association of Physics Teachers

AAS   American Astronomical Society

ACA   American Crystallographic Association

AGU   American Geophysical Union
AIP   American Institute of Physics
ANS   American Nuclear Society
APS   American Physical Society
ASA   Acoustical Society of America
ASMS   American Society for Mass Spectrometry
AVS   American Vacuum Society
AWIS   Association for Women in Science
DOD   Department of Defense
DOE   Department of Energy
EPA   Environmental Protection Agency
NASA   National Aeronautics and Space Administration
NIST   National Institute of Standards and Technology
NOAA   National Oceanic and Atmospheric Administration
NSBP   National Society of Black Physicists
NSF   National Science Foundation
NTIA   National Telecommunications and Information Administration
OSA   Optical Society of America
SAS   Society for Applied Spectroscopy
SETI   Search for Extraterrestrial Intelligence
SoR   Society of Rheology

INDEX

David E. Newton has published extensively on physics and other science subjects. He is the award-winning author of numerous books, articles, and scholarly publications, including *Chemistry* (the first volume in the Oryx Frontiers of Science Series), *The Ozone Dilemma, Encyclopedia of Cryptology,* and *Global Warming.* Newton received a doctorate in science education from Harvard University.